# THE ENDURING WILD

# THE ENDURING WILD

---

## A JOURNEY INTO CALIFORNIA'S PUBLIC LANDS

---

### JOSH JACKSON

*Berkeley, California*

Copyright © 2025 by Josh Jackson

All rights reserved. Except for brief passages quoted in a review, no portion of this work may be reproduced or transmitted in any form or by any means, electronic or mechanical, including photocopying and recording, or by any information storage or retrieval system, or be used in training generative artificial intelligence (AI) technologies or developing machine-learning language models, without permission in writing from Heyday.

Library of Congress Cataloging-in-Publication Data
Library of Congress Cataloging-in-Publication Data

Names: Jackson, Josh, 1979- author.
Title: The enduring wild : a journey into California's public lands / Josh
  Jackson.
Description: Berkeley, California : Heyday, [2025] | Includes
  bibliographical references.
Identifiers: LCCN 2024044470 (print) | LCCN 2024044471 (ebook) | ISBN
  9781597146753 (hardcover) | ISBN 9781597146760 (epub)
Subjects: LCSH: Wilderness areas--California. | Public lands--California. |
  California--Description and travel. | Jackson, Josh,
  1979---Travel--California.
Classification: LCC QH76.5.C2 J34 2025  (print) | LCC QH76.5.C2  (ebook) |
  DDC 508.794--dc23/eng/20241227
LC record available at https://lccn.loc.gov/2024044470
LC ebook record available at https://lccn.loc.gov/2024044471


Cover Art: Josh Jackson
Cover Design: Noah Smith
Interior Design/Typesetting: *the*BookDesigners
Interior Maps/Illustrations: Rebekah Nolan

Published by Heyday
P.O. Box 9145, Berkeley, California 94709
(510) 549-3564
heydaybooks.com

Printed in China by Imago

10  9  8  7  6  5  4  3  2  1

*For Kari, Stella, Leo, and Vivian, my beginning and end*

# CONTENTS

Introduction . . . . . . . . . . . . . . . . . . . . . . . . . . . . . . . . . . . . . . . . . . . . . . . . . . . . *1*

The Mojave: A Pilgrimage . . . . . . . . . . . . . . . . . . . . . . . . . . . . . . . . . . . . . *29*

Carrizo Plain: Place Attachment . . . . . . . . . . . . . . . . . . . . . . . . . . . . . . . *57*

Borderlands: The Story of Hunting . . . . . . . . . . . . . . . . . . . . . . . . . . . . . *81*

Eastern Sierra: There's Gold in the Hills . . . . . . . . . . . . . . . . . . . . . . . . *111*

Berryessa: The Radical Center . . . . . . . . . . . . . . . . . . . . . . . . . . . . . . . . *141*

King Range: Reciprocity . . . . . . . . . . . . . . . . . . . . . . . . . . . . . . . . . . . . . *161*

Wilderness, Reimagined: A Walk in the River . . . . . . . . . . . . . . . . . . . *189*

Epilogue . . . . . . . . . . . . . . . . . . . . . . . . . . . . . . . . . . . . . . . . . . . . . . . . . . *215*

*Acknowledgments* . . . . . . . . . . . . . . . . . . . . . . . . . . . . . . . . . . . . . . . . . . . *223*

*Guide for Exploration* . . . . . . . . . . . . . . . . . . . . . . . . . . . . . . . . . . . . . . . *225*

*Selected Bibliography* . . . . . . . . . . . . . . . . . . . . . . . . . . . . . . . . . . . . . . . *239*

*About the Author* . . . . . . . . . . . . . . . . . . . . . . . . . . . . . . . . . . . . . . . . . . *245*

*A Note on Type* . . . . . . . . . . . . . . . . . . . . . . . . . . . . . . . . . . . . . . . . . . . . *247*

*American conservation is, I fear, still
concerned for the most part with show pieces.*

—ALDO LEOPOLD, 1948

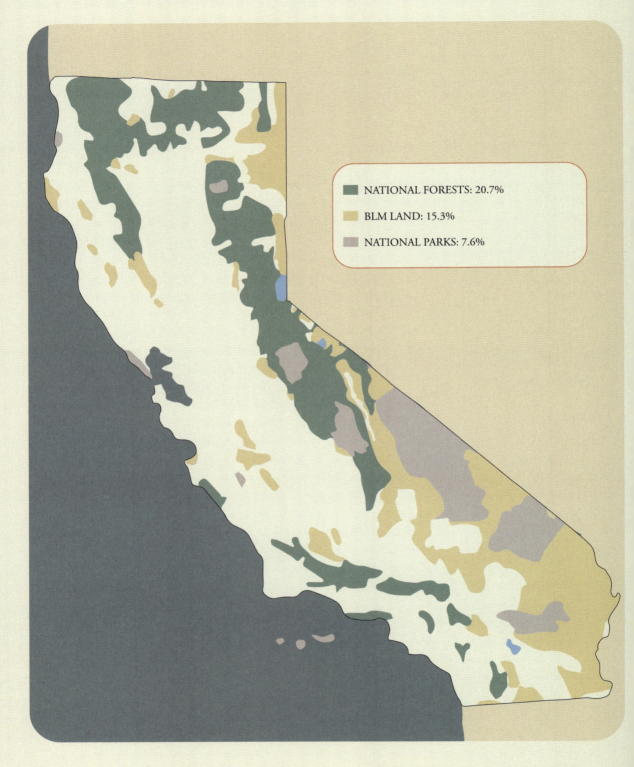

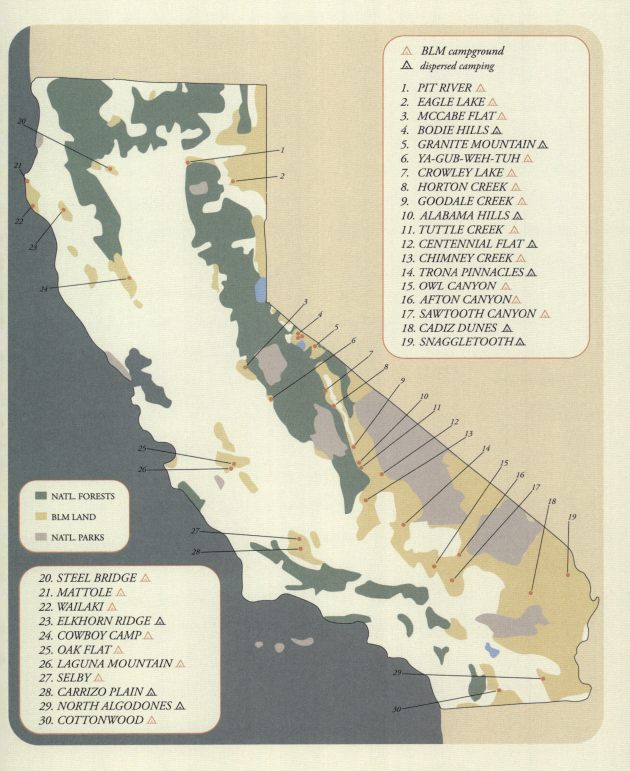

# INTRODUCTION

*I was simply looking* for a place to camp. That's how this whole thing started.

It was January 2015. We had a newborn at home, and my partner needed some extended rest, some quiet within the crowded rooms of our two-bedroom house. We had been hunkering down as the pregnancy deepened, avoiding long trips, and surrendering happily to those last hallowed months we had as a family of four. After our family grew to five, my older two children and I decided to hit the road for a few nights, longing for our own kind of quiet and for those bright stars against a dark sky.

Our two-year-old son, Leo, was just emerging from a life marked by colic, and his personality was brimming with newfound independence. But as it goes for most parents navigating the stormy unpredictability of toddlers, we still weren't getting out much. Plus, our daughter Stella, five at the time and a kindergartener, was in her first year at "real school," as she called it, so our family's schedule was also dictated by an orderly calendar that didn't leave much room for traveling or adventure. The writer Robert Macfarlane wrote about this sense of confinement: "Anyone who lives in a city will know the feeling of having been there too long. The gorge-vision that the streets imprint on us, the sense of blockage, the longing for surfaces other than glass, brick, concrete and tarmac."

After our youngest daughter, Vivian, was born in late November 2014, an event that coincided with the holidays and family visits and more hunkering down, I was eager to find some narrow country roads that might lead to anywhere with more earth than concrete. I searched and searched for available campground sites, but everything within a few hours of Los Angeles was booked solid. We were dismayed.

In a last-ditch effort to find something, I reached out to my friend Justin, the only human I knew who had a special knack for making campsites appear out of thin air. I explained my predicament and he said, "What about BLM land? I don't think you need reservations."

To which I replied, "Where is this magical place you speak of?"

"Out in the desert." He said this, half perplexed, almost like a question. As if to say, *I think it's out in the desert somewhere?*

Even as an avid outdoor enthusiast, I had never heard of "BLM land" before that conversation. After hours of sifting through the BLM website, which felt like going back in time to the pixelated domains of dial-up, I found a place called the Trona Pinnacles, located in the middle of the Mojave Desert, a breezy 171 miles from my home in Los Angeles. We loaded up our Honda Element and headed east, not quite knowing what to expect. But we were doing it! We were leaving our grand metropolis and heading somewhere *out there*, away from the hum of traffic, overhead helicopters, and the constant flicker of artificial light. I wasn't even concerned about what the Trona Pinnacles looked like; the act of leaving was more important than whatever greeted us upon arrival.

As Los Angeles shrank in the rearview mirror, ever so slowly, and the landscape flashing by transformed from skyscrapers to industrial buildings to quiet neighborhoods to suburbia and finally to the open vistas of the desert, a sense of lightness settled over me. When we finally arrived at the Trona Pinnacles entrance off Highway 178, renewal was already threading its way through my overstimulated mind. We pulled off the two-lane road and parked near a weatherworn sign about the geological formations we were headed toward. The Mojave air was clean and cold, and I couldn't help but draw in a deep breath.

While I scoped out the driving conditions that would take us another six miles across the salt flats, the kids were already out of the car, zigzagging around saltbush and desert holly, pointing at this and that, yelling about trains and lizards. And then, with Leo in my lap helping me steer down the empty road, and Stella proudly riding shotgun, we bounced and fishtailed and hollered our way to our destination.

What we found was a fantastical scene: hundreds of tufa spires protruded like drip-style sand castles out of the wide-open desert floor that extended for miles in every direction. It was as if we had magically transported ourselves to another planet. We spent two days wandering around the spires, climbing boulders, and pretending the bouncy salt flats were a giant trampoline. As we departed, I felt the clearest I had in a good long while. And the kids had that serene, post-exploration glow about them, their faces untroubled and their bodies relaxed after going so hard. They were fast asleep before we even reached the highway.

The trip served as my introduction to BLM lands and became the catalyst for a journey that quickly turned into a fascination and eventually an obsession. The Trona Pinnacles fulfilled all the things I longed for at the time—solitude, reprieve, bare earth, and no-reservation camping. They also rekindled my desire for adventure. So, I did what any parent of three kids under the age of six does when the desire to hit the open road beckons: I ordered books.

Since our travel options were newborn-limited, I turned to the source of learning that had consistently guided me. Books have always been a through line in my life, from the tales that captivated me in grade school to the novels and memoirs that expanded my worldview and piqued my curiosity into adulthood. If I couldn't get out there as much as I wanted for a few months or years, I decided to spend that time researching, reading, and learning everything I could on the subject of BLM lands. Where could I find them? What kinds of landscapes did they contain? And what exactly were they?

I started with the most straightforward titles, which ended up being *These American Lands* by Dyan Zaslowsky (1986) and *America's Public Lands* by Randall K. Wilson (2014). Both books offer concise histories and descriptions of the different government agencies that manage public land, and with their help I learned enough to begin pouring a foundation of understanding.

The BLM—short for the Bureau of Land Management—was established in 1946, when the Department of the Interior merged the General Land Office with the Grazing Service. Today, the BLM is one of four federal agencies that manage public land across the United States. The others are the National Park Service (managing 84 million acres), the US Forest Service (193 million acres), and the US Fish and Wildlife Service (96 million acres). As for the BLM, it manages an astonishing 245 million acres across the western states and Alaska. I was dumbfounded by this number. Two hundred and forty-five million acres? And then another question emerged: How on earth had I never heard of these lands?

Even the idea of "public lands" wasn't a concept I fully understood. But as I learned more about them, I came to see public lands as something miraculous. When you consider our national zeal for owning private property, it is astounding we have set aside 618 million acres of federal land for the public. These are areas

Trona Pinnacles

of land and water that are *owned collectively by the citizens* and managed by the federal government. The concept is worth reiterating: These lands are our common ground, a gift of seismic proportions that belongs to all of us. No matter your color, creed, or class, and even if you've never signed your name to a deed, you are a landowner. We are all landowners. Coming to terms with this endowment was a revelation.

Those first two books on the subject introduced me to several other books (the great inevitability of reading, as if the books themselves have formed a secret alliance to support one another), which then led me to dozens of other titles across various perspectives and time periods. I read histories, biographies, nature atlases, travelogues, memoirs, and guidebooks written by a wide variety of thinkers and recreationists from different backgrounds.

Using one of Stella's largest sheets of watercolor paper, I started to keep track of each title and author, the big ideas they represented, and how each one led to the next. Before long, I had an interconnected network of lines, arrows, question marks, and dates, and the paper looked like one of those walls on which, at least in the movies, police investigators track criminals by piecing together information, clues, and pictures with thumbtacks and string. The more I delved into the history, key figures, and laws shaping BLM lands, the better I understood what sets the bureau's holdings apart from those of other federal land agencies. It was these ideas that eventually led me to write this book.

The first distinction that makes the BLM lands so unique is how they came to be in the federal domain. Many historians refer to them as the "leftover lands" simply because nobody seemed very interested in acquiring them. They weren't suitable for homesteading, even when the government was offering large tracts through the Homestead Act. They weren't profitable enough to be purchased by developers, land barons, railroads, or corporations. And even the other federal agencies that manage public land passed them over. They were considered too unattractive for the National Park Service, too bare for the Forest Service, and too lifeless for the Fish and Wildlife Service.

In the end, these millions of leftover acres were entrusted to the Bureau of Land Management. The only thing they had in common was that no one else wanted them, a fact that explains why it's so hard to describe what kinds of landscapes BLM land includes. Having experienced the delight and wonder of the tufa-spired desert, I started to wonder if maybe the Trona Pinnacles were an exception; maybe

RIGHT: Cache Creek Natural Area

the rest of the BLM's holdings were as plain, denuded, and inanimate as their origins suggested they might be.

Of course, there's a deeper story behind how these lands became part of the federal domain, and it likely won't surprise you to learn that it's an undeniably ugly one. It didn't take long for me to discover that these lands were systemically stolen from the hands of the Native Americans who had stewarded them for millennia before colonialism in different forms devastated their tribes. Through invasions, plagues, violence, coercion, bribery, war, and lopsided deals with the United States government, Indigenous groups and their ways of life were nearly obliterated. Behind every layer of history I peeled back was a heartbreaking story of loss.

The second distinction that sets BLM lands apart lies in how these leftover lands were perceived and treated. Once they were deemed unsuitable and/or unprofitable for homesteaders, developers, and other federal land agencies, their fate seemed inconsequential. As a result of this perceived inferiority, vast areas of those lands were then severely degraded through both overgrazing and unchecked extraction of coal, gas, oil, and various minerals. Our BLM lands, while managed by a federal agency, were not always protected by it.

From its inception in 1946 until the passage of the Federal Land Policy and Management Act (FLPMA) in 1976, the BLM primarily managed lands for grazing and extraction. The agency was so synonymous with those two uses, it was nicknamed the "Bureau of Livestock and Mining" by critics who thought the agency emphasized these activities at the expense of other land uses and environmental considerations. When FLPMA passed by an overwhelming 78–11 vote in the Senate, it called for a multi-use mandate that directed the BLM to more thoughtfully balance grazing and extraction with conservation and recreation.

In the wake of FLPMA and other environmental laws passed during the 1960s and '70s, however, many members of the ranching and extraction industries (and the politicians they supported) were angered by what they viewed as government overreach. The reality was that their unfettered access to BLM land was suddenly being curtailed by FLPMA's mandate to consider other uses. This discontent culminated in the Sagebrush Rebellion of the 1970s and '80s, which demanded reduced regulation and oversight. Although the rebellion fizzled out after President Reagan's election, the ideology has resurfaced in various forms ever since, including, most notably, during the 2016 standoff, fronted by anti-government militia leader Ammon Bundy, at the Malheur National Wildlife Refuge in Oregon. The

LEFT: Pit River

Trinidad Head

continued influence of the Sagebrush Rebellion prioritizes reducing federal control to benefit individuals and corporations, often at the expense of the public interest and the protection of public lands.

Despite FLPMA's mandate for a more balanced approach to multi-use land management, the grave imbalance between conservation and industry still exists almost five decades later. The activist group Public Employees for Environmental Responsibility (PEER) reported in 2024 that a quarter of all grazing lands fail to meet BLM's own rangeland health standards. At the same time, extraction industries continue to threaten biodiversity, to fragment wildlife corridors, and to pollute air, water, and soil. Of the 245 million acres managed by the BLM, only 37 million acres (15 percent) have been set aside for conservation. While I fully support the multiple-use mandate for BLM lands, and I appreciate the role that cattle, natural gas, oil, and certain minerals play in our everyday lives, I still see the pendulum swinging too far toward industry.

The fortunate 15 percent that receive higher levels of protection, known as National Conservation Lands, include National Monuments, Conservation Areas, Scenic Areas, Recreation Areas, Wild and Scenic Rivers, National Trails, Wilderness Areas, and Wilderness Study Areas. As stated at the establishment of the National Conservation Lands in 2000, the purpose of this designation is to "conserve, protect, enhance, and manage public lands for the benefit and enjoyment of present and future generations." They are protected through congressional legislation, BLM management decisions, and presidential proclamations under the Antiquities Act.

Established by Theodore Roosevelt in 1906, the Antiquities Act grants presidents the authority to designate natural, historical, and cultural areas as National Monuments for permanent protection. In fact, many of our beloved National Parks were originally set aside as National Monuments, including the Grand Canyon and Joshua Tree, and almost every single president since Roosevelt has used the act to establish a new monument or expand an existing one.

Yet, even protected BLM lands face threats. On December 4, 2017, President Trump became just the third president to *reduce* the size of a National Monument, cutting protections on 2 million acres of BLM land in both Bears Ears and Grand Staircase–Escalante, in Utah. As a result, land sacred to the Navajo, Ute, Hopi, and Zuni tribes was suddenly opened to new mining, fracking, and drilling claims.

Learning about such land-use imbalances and the fragility of these supposedly protected areas was a sobering wake-up call for me, and it intensified my desire to

RIGHT: Point Arena–Stornetta

Carrizo Gorge Wilderness

experience these lands firsthand. But despite my growing knowledge, I still struggled to find certain details about BLM land within California, including how curious visitors could access them and what we might find there. The lack of personal narratives made these lands feel distant and disconnected, as though they existed only in the dry facts of government websites. Stories have a way of breathing life into places, helping us connect on a deeper level, but in my BLM research, such accounts were either absent or merely footnotes in major narratives about the West. It was no wonder I hadn't heard of these lands before.

Fast-forward to April 2020. I was sitting on the front porch with my wife, Kari, sipping coffee while the usual city noises went eerily quiet. Songbirds now took center stage, as if their harmonious volume had been methodically dialed up each day since the start of the pandemic. Changes to our lives had come fast and furious. The kids now greeted their teachers from our dining table, and just as my small furniture business was about to celebrate its tenth anniversary, our "non-essential" retail store shut down, and custom orders were drying up. Since our trip to the Trona Pinnacles, my three children had grown five years older, and instead of having three kids under the age of six, we were on the brink of having three kids over the age of six.

Kari and I sat on the porch and discussed so many things that day: the unknown monster called COVID-19, our jobs (she was an ER nurse at the time), the challenges of parenting in this new reality, and what the hell we were going to do with our lives. It was a "State of the Family" type of conversation. If there ever were a time to pivot, to make some kind of bold leap into the unknown, this was it. After getting more coffee, we sat in thoughtful silence, watching a northern mockingbird in a nearby cypress. Every morning at the same time, he sang from the very top of the highest branch, perched like a star atop a Christmas tree.

Then out of nowhere, Kari said, "Why don't you start going to see these BLM lands? You've been talking about it forever. I think this is the time. We'll be good."

And that was it, the catalyst I needed. As our mockingbird sang his songs and the kids went to school on laptops, my journey to find and experience these lands had begun. We agreed to take it one trip at a time, between Kari's hospital shifts and orders coming in at my day job. We would dive into the shallow waters of our savings account to make it work. The objective would be to systematically visit as much BLM land in California as I could access with our two-wheel-drive family van.

I had almost no idea what lay ahead, but I wanted to find out. If these so-called leftover lands had a story to tell, I wanted to play a small part in telling it. So, with camera in hand and notepad in pocket, I hit the road. Again and again and again, over the course of forty-two months and thirty-two trips, I traversed the Golden State, from Mexico to Oregon, Nevada to the Pacific. I have taken in these places slowly, step by step, attuning myself to their rhythms, and walking four hundred miles along the way.

King Range National Conservation Area

During this time, I have also met and listened to the incredible people who work and care for these lands, and whether they were nonprofit workers, members of Native communities, policy advocates, volunteers, or dedicated civil servants at the BLM, all offered me invaluable insights that have led to my growing intimacy with these landscapes. Even as I slowly filled the story void, another set of questions emerged, only now finding the answers seemed even more consequential than before. My initial fascination with exploring new landscapes had deepened into a commitment to protecting all that I'd experienced. If these precarious places go unseen and unspoken, who will notice when the subtle beauties of desert, sagebrush, grasslands, and remote mountains slip away under the pressure to turn places into profits? In other words, how can we protect what we don't know?

Despite their origin as "leftover" lands and the severe overgrazing, unchecked development, and resource extraction they've faced since their inception, BLM lands have endured. Their resilience lies in their ability to withstand adversity, reminding us that even lands once overlooked and undervalued possess a profound strength and beauty worth fighting for.

If we are going to expand protections on our most vulnerable public lands, preserving them for future generations, we will need a much broader coalition of people who love and speak up for them in the present. My desire for this book is that it will highlight what I have seen and what I have learned by telling the lands' stories through words, images, maps, and illustrations. I hope this can be a starting point for many of you to have your own experiences on these lands, and I hope you'll find your own role in helping to protect them.

For the wild—in us, out there, and down the street.

*Josh Jackson*
JULY 2024

LEFT: Virginia Creek

Bodie Hills

Cadiz Dunes Wilderness

Alabama Hills National Scenic Area

Granite Mountain Wilderness

Carrizo Plain National Monument

Owens Peak Wilderness

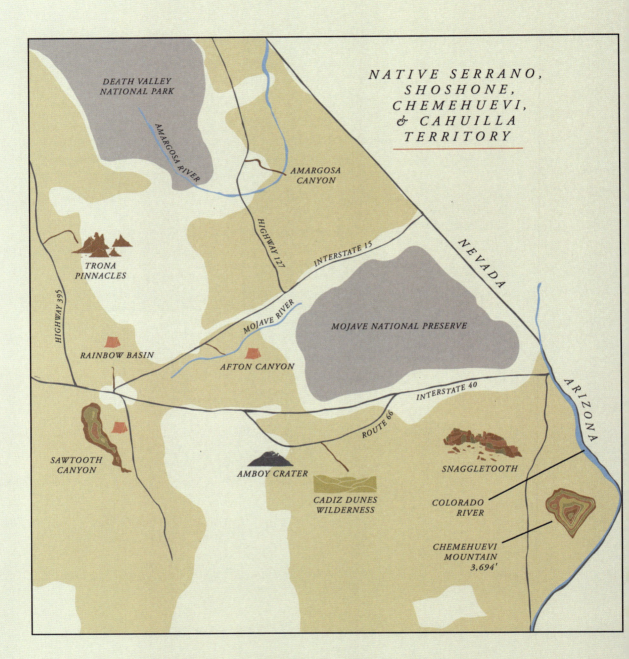

MOJAVE DESERT

# THE MOJAVE: *A Pilgrimage*

*A large-scale map* of California hangs on the wall in our living room, practically filling up the entire space between the floor and ceiling. Standing in front of it, my eyes are level with Mount Shasta, and my knees land near my home in Los Angeles. Printed in 1989, it was produced to highlight the different federal agencies that manage public land across California. Each agency's holdings are marked by a different color, making the map a beautiful medley of hues and shapes that fit together like a puzzle. The vast network of National Forest lands are shaded green. The National Parks' are purple. The isolated pockets of the Wildlife Refuges are blue. Everything washed in white is privately owned land, which fills up the space around cities, along the Pacific coast, and in the San Joaquin and Sacramento Valleys, which run between Bakersfield and Redding. The map is exhaustive, both in capturing the full breadth and scale of the Golden State and also in highlighting with precision the rivers, mountains, deserts, and forests that make up the sum of its landscapes.

My preoccupation is with the yellow-shaded areas of the map, which are the public lands managed by the Bureau of Land Management. If you were to examine the map up close, you would notice that the smaller parcels take on various shapes and sizes and are scattered all around the state, often existing on the edges of National Forests or as islands in a sea of private land. But if you were to take a step back and look at the map in its entirety, it is abundantly clear where the majority of BLM land exists: the Mojave and Colorado Deserts of Southern California. Showcased in dense yellow clusters, these areas loom large on the map, filling much of the space along California's borders with Arizona and Nevada, as well as most of the desert land outside of the area's National Parks. The odds are quite good that if you parachuted into the Southern California desert, you'd touch down on soil managed by the BLM.

Centennial Flats

Teddy-bear cholla

Together, the vast expanses of BLM land in the desert plus the smaller tracts interspersed through the state add up to fifteen million acres of land, more than double the combined acreage of the state's nine National Parks.

Over the course of five years, I had developed an ardent preoccupation with these often-overlooked landscapes. What started with that weekend trip to the Trona Pinnacles in 2015 had become something more significant as I learned about the threats these lands face. It was like reading a mystery novel full of heroes and villains, surprise and intrigue. Only this novel had no ending—the fate of these landscapes still under threat, and the multi-use mandate still out of balance.

In 1948, renowned naturalist Aldo Leopold, discussing conservation in his home state of Wisconsin, wrote: "American conservation is, I fear, still concerned for the most part with show pieces. We have not yet learned to think in terms of small cogs and wheels." During a time when Americans were primarily interested

Centennial Flats

in protecting the sublime beauty of the National Parks, Leopold called for a broader understanding of conservation. I began thinking of BLM lands as the "cogs and wheels" of modern conservation. Although not as prominent as the "show pieces," these lands are equally important, serving as vital pieces of the federal public land puzzle. Protecting more of these lands would enhance biodiversity, help mitigate the effects of climate change, provide safe havens for threatened species, and serve as wildlife corridors connecting other federal lands.

The map, hanging in my living room like a beacon, served as my starting point. Looking at it was exhilarating, a reminder that I would eventually visit all the yellow-shaded areas spread throughout California. It felt like unwrapping a gift of immeasurable proportions, with millions of acres of unknown and unexplored lands just on the horizon. However, the map was also intimidating. I had almost no idea what kinds of landscapes I would be encountering. With a pencil in one

hand and a brand-new notepad in the other, I stood before the map and wrote down every name and natural landmark labeled within the yellow-shaded areas. I also called the BLM's state headquarters in Sacramento and ordered nearly every paper map they had for each district in California. Two weeks later, sixty-six maps arrived in a large cardboard box, and I spread them out on my dining table, organizing them district by district until I had connected them all.

My first area of interest was the Mojave Desert, and I focused my attention on finding campgrounds and hiking trails—my preferred ways to connect with nature. I painstakingly searched the BLM's online resources dedicated to the Mojave Desert, gathering as much information as I could. When those sources didn't provide everything I needed, I turned to blogs, summit reports, online forums, mapping tools, and hiking apps. By the time I had finished my research, my notepad was filled with enough locations and landmarks to keep me occupied in the desert for a good long while.

For my inaugural desert trip, I chose a place called the Rainbow Basin Natural Area, a small two-thousand-acre swath of BLM land on the western half of the Mojave, 3,785 feet above sea level.

---

A series of long gravel roads with heavy washboard ridges brought me to the Owl Canyon Campground just as daylight crept slowly across the horizon. The campground was empty aside from a modern RV parked at site number 2 and the owners' barking dog, who welcomed me as I passed by. I felt a nervous excitement as I deposited my nominal camping fee into an envelope and picked a site for the weekend. My goal was to see as much of the natural area as I could in three days, and I couldn't wait to dive in.

By mid-morning, I was three and a half miles into my first hike and standing alone on a ridgeline. The winter clouds had only recently moved on, allowing the sun to make its grand entrance and offering me uninterrupted views. Hovering in the distant north were bare mountain ranges stacked and silhouetted one on top of another, their formations and elevations stretching as far as my eyes could take me. Open wastelands of brush and a network of neglected off-highway-vehicle trails crisscrossed the landscape, mirroring the plane contrails above. Rocky outcrops jutted from the nearby hillsides, and a scattered stand of Joshua trees with their bright-green, bayonet-shaped leaves provided the only contrast to the dull colors of the desert floor. Some elder Joshuas lay strung out along the ground, their trunks toppled and in various stages of decay.

It was getting warmer and I was only halfway through my self-guided hiking loop, which had thus far brought me up through a waterless slot canyon of various widths and pitches. Underneath my backpack my shirt was soaked in sweat, and I was ready for a break. I methodically checked the ground for camouflaged rattlesnakes and then dropped my pack and my tired body on the soft sand. There was no wind, no birdsong, not even distant noise.

Looking at the arid landscape, I was awestruck by the sheer scale of barrenness. Words describing the scene spun through my mind like a thesaurus entry. Bleak. Austere. Infertile. Empty. Forsaken. Everything appeared to be running on repeat: rocks, sand, shrubs, and mountains, an endless array of the same muted colors and interchangeable rock formations. As I sat with my back against the craggy bark of a Joshua tree, I couldn't help but feel underwhelmed. But I also couldn't shake the sense that I was missing something.

Before embarking on this initial journey to the parched country, I had read a memoir called *Refuge* by Terry Tempest Williams. In it, the author, writing about her mother's struggle with cancer, gracefully weaves that narrative together with the environmental issues threatening the Great Salt Lake and the high desert that surrounds it—a landscape Williams had intimately known her entire life. As I now stared at the seemingly lifeless expanse before me, I was drawn back to the evocative words from *Refuge* that I had copied into my notepad: "If the desert is holy, it is because it is a forgotten place that allows us to remember the sacred. Perhaps that is why every pilgrimage to the desert is a pilgrimage to the self. There is no place to hide and so we are found."

I sat for a long time under the shade of the Joshua, letting Williams's sentiment bounce around my mind, flipping it up in the air, questioning it. Was there really something holy about this desert? Something sacred? Was a pilgrimage to the desert really a pilgrimage to the self? For the rest of the weekend, as I hiked around Rainbow Basin, her words rang out like a riddle, one I found difficult to answer.

Back at home, I stood before the map once again, feeling deflated. The scenery of the BLM lands in Rainbow Basin had disappointed me, and I wondered if the

rest of the desert I intended to visit would make me feel the same way. But over the following weeks, as I returned to work in my furniture studio, Williams's poetic language stayed with me, like a whisper, like a gentle tap on the shoulder. I kept coming back to that word *pilgrimage*. "Perhaps that is why every pilgrimage to the desert is a pilgrimage to the self," she wrote. I started to wonder if Williams's words were a benevolent invitation to continue my journey by looking inward.

The more I considered treating subsequent trips as a pilgrimage, the more I thought about hiking versus walking. I had always prided myself on being a hiker, having logged almost a thousand miles on backcountry trails, but hiking has a different set of goals and priorities than walking—calculated objectives such as speed, mileage, elevation gain, and reaching a destination. But the idea of walking felt like something else. I started to think of walking as a deliberate act of slowing down, of letting go of an agenda or a purpose. What if walking, in its most elemental form, was about noticing?

Maybe, through the contemplative steps of a pilgrim, I could unlock the holiness of the desert. And maybe something in myself too.

---

A month later, I was on the road again, headed straight for the interior of the Mojave. I still had more questions than answers, but I also had an intention to put the act of walking at the center of my journey. For my second trip, I chose a BLM-managed area called the Amargosa Canyon, which lies south of Death Valley National Park and is home to perennial streams, endangered species, and the federally designated Amargosa Wild and Scenic River.

Following my decision to focus on noticing rather than achieving, my approach to planning for the Amargosa underwent a pivotal shift. In Rainbow Basin, I had been focused on how I could *use* the land through hiking and camping; for this trip, however, I aimed for a more comprehensive understanding of the land itself. The Amargosa Conservancy, a local nonprofit that advocates for the region and provides stewardship opportunities, became an invaluable resource, and using their website as a starting point, I was able to learn about the area's rich human history, diverse flora and fauna, and intricate ecosystems.

Rainbow Basin Natural Area

The most accessible entry point to the Amargosa is via the China Ranch Date Farm, a privately owned business bordering the BLM land. After driving down an empty two-lane highway for an hour and then navigating a series of remote gravel roads, I was surprised to suddenly drive past old barns, manicured rows of date trees, and a gift shop where travelers can purchase souvenirs and date milkshakes. I parked in a small lot at the back of the date farm, where I was greeted by informational signs, picnic tables, and a vault toilet, all of which were a welcome sight after my four-hour road trip.

A couple who looked to be in their late fifties were standing near the back of their truck camper, spooning soup out of ceramic coffee mugs. She wore an oversized button-up flannel, and he wore blue jeans and a trucker cap. They appeared as weathered as their old pickup. Just as I opened the door, the woman called over.

"Hey!" she shouted with a smile, as if I were arriving right on cue for a moment I didn't know was coming. "So what are you doing here?"

"Hi," I replied, somewhat startled. "I'm here to see water in the desert."

"It's out that way," she said, pointing toward a well-trodden trail. "Hard to believe it sitting here, but it's out there. And there is a waterfall too."

"A waterfall?"

"Yeah, a waterfall!" she shouted with a sudden burst of liveliness, her eyebrows raised as high as her voice. "It's tiny, but it's a real live waterfall! You have to hike a ways to get there, though."

"Well, that's exactly what I'm here for," I said warmly, trying to match her cheerful disposition. I didn't mention I was on a pilgrimage. Nor did I use the word holiness.

I loaded my backpack while we chatted some more, the talk mostly revolving around the couple's recent camping adventures up the road in Death Valley. The woman went on about the cold temperatures in the higher elevations, while her husband ate his soup without even looking up. It started to make sense why she was so eager for conversation.

"Now, Death Valley. *That* place is amazing." She said this slowly, with a heavy emphasis on the word "that," which I took as an implication that the canyon I was about to enter wasn't as impressive.

By the time I finished packing, the couple had wandered off to the date farm and it was a few minutes past noon. My backpack was stuffed with extra clothes, books, a zoom lens, and enough food and water to last several days, just in case things went awry. A hip pack held my phone, a wilderness whistle, a pencil, and

RIGHT: Cadiz Dunes Wilderness

my notepad. Binoculars and a camera hung from my shoulder. As I set off, I realized how ridiculous I must have looked, fit to be embarking on a long expedition when, in fact, I needed to be back by nightfall to find a place to camp.

I stepped onto the trail with great anticipation, but almost immediately I noticed the pull of my natural inclination to conquer each mile of available trail and photograph as much of the canyon as possible. I had started calculating whether or not I could cover a dozen miles before dark and what routes I would take. And then I remembered. I was there to *walk*. I pulled out my notepad and wrote the words *slow down*.

The sky was heavy with cloud cover, a permanent gray in various shades of dark and light. It felt as if dusk had arrived too early, creating a moody ambiance. The temperature hovered in the low 50s, and I was grateful to be layered up. The wide, developed trail followed Willow Creek, which carried a shallow but steady flow of water. A host of small birds darted back and forth between the willows and the creekbed, but they were too sly and quick to identify. The least Bell's vireo, a small songbird and one of several endangered species that rely on the canyon, would be arriving in a few months for breeding season, one of some two hundred fifty avian species that exists within the Amargosa, including both local and migratory populations. The birds' fluttering and occasional chirping were the only sounds that broke the desert silence.

The area around what we now call Amargosa Canyon is the native homeland of the Timbisha Shoshone, who have inhabited the region for at least a thousand years. The name Timbisha comes from the red ochre found in the mountains near Death Valley, and it was used as a paint for protection, healing, and spiritual strength. The people call their ancestral homeland Tüpippüh, which encompasses the canyons, mountains, valleys, dunes, and waterways throughout the entire Amargosa basin and beyond. After a long and arduous battle, the tribe finally received federal recognition in 1983, and under the Timbisha Homeland Act of 2000, 7,754 acres of their land was returned by the US government. In a preface to the "Draft Secretarial Report to Congress," tribal elder Pauline Esteves beautifully captured the story: "Our people have always lived here. The Creator, Appü, placed us here at the beginning of time. . . . We never gave up. The Timbisha people have lived in our Homeland forever and we will live here forever. We were taught that we don't end. We are part of our Homeland and it is part of us. We are people of the land. We don't break away from what is part of us."

LEFT: Amboy Crater

As the trail veered from the creek and the canopy slowly faded away, I began to see signs of past resource extraction. In the early twentieth century, before the BLM was formed, borax was mined here and transported out to the rest of the country via the Tonopah and Tidewater Railroad. The numerous abandoned borax mines carved into nearby hillsides and the decrepit railroad tracks running haphazardly through the canyon served as stark reminders of how these lands were viewed by the settlers who claimed the tribe's homeland for their own uses. Today, thanks to the Wild and Scenic River designation, the canyon itself is protected against future extraction.

Past the mines, the views started opening up to reveal the canyon in more vivid detail. The recent rains and overcast sky made for striking colors on the tallest mountain, which rose over the canyon like a monolith some seven hundred feet above the valley floor. I couldn't take my eyes off it. It was a rainbow of diverse sedimentary rock from top to bottom, its fractures, rifts, and cracks running horizontally like compressed lines on a topography map. From the faded black rock beds all the way up to the top of the ridge unfurled a patchy concoction of colors: chocolate brown, crimson, burnt orange, hazel, and bands of white and pink. I'm not sure I've ever seen a mountain so vibrant, as if the entire formation had gone through a rock tumbler and then been polished with a matte oil finish.

Creosote lives here in small pockets, not nearly as densely as it does in other parts of the Mojave, but what the plants here lack in quantity, they make up for in size: Some of their crowns grew up past my eye level and spread out twenty-five feet in diameter. Other types of plant life were more abundant, including the low-growing seepweed, arrowweed, longleaf jointfir, desert holly, saltbush, and white bursage, which together with the creosote completed the desert diorama. I stopped to inspect each new plant I encountered, studying their roots, branches, and leaves. Seeing them up close, I began to get a sense of their uniqueness, their subtle colors, their importance in the greater ecosystem. The complex root system of the creosote, for instance, offers an underground den to the kangaroo rat, while the seeds of the white bursage provide meals to birds and black-tailed jackrabbits.

I also spotted, hanging in heavy clumps, mostly in the wild branches of the thorny mesquite, the guileful desert mistletoe, otherwise known as the tree thief. On

## RAINBOW MOUNTAIN
*Amargosa Canyon*

first glance, the mistletoe seemed to be using the branches of the mesquite to climb closer to the sun, and doing it so thoroughly that it was hard to differentiate which branch was from which species. The reality of the matter—via a relationship as old as the Holocene epoch—was even more mysterious: The seeds from the pink berries of the mistletoe actually germinate on the very bark of the host tree. The mistletoe then sends out haustoria—specialized rootlike structures—into the microscopic spaces between the tree's cells, withdrawing water and nutrients from the mesquite. It's a one-sided embrace that slowly saps the host tree's strength, leaving it vulnerable to the harsh whims of the desert.

As I edged closer to the lookout above the river, where the canyon makes a wide turn from the southeast to the southwest, I was stopped in my tracks by what I saw off the trail, down along the canyon floor. A lone coyote was standing perfectly still and looking up at me. I paused and slowly raised my binoculars until we were eye to eye, its gaze studying my every move. I stared into its eyes, a hypnotizing blend of yellow and brown, and I felt my heart rate pick up even though I knew the coyote wasn't a threat. The Timbisha word for coyote is *e-jah*, and the animal plays a significant role in the tribe's spiritual ideology and storytelling, acting as an important messenger. I wondered what kind of message this coyote might have for me as I visited these lands. After a minute, it turned and sauntered down toward the water, disappearing into a gully and out of sight.

Standing at the canyon's lookout, I took in the majesty of the scene. The canyon is immense, running hundreds of feet wide in some places, with vertical walls ten to twenty feet high on both sides, carved out by a millennia of flash floods. The tanned walls look like smooth dried mud, something akin to the outside of a massive wasp nest, complete with what looked like entrance and exit doors. Along the riverbank, cattails are stacked up uniformly, perfectly spaced apart, like a thousand little fingers reaching from the shallow banks. Clumps of salt grasses run up to the base of the canyon walls, intermingling with willow and invasive tamarisk. Between the rainbow stripes of the looming mountain and the green, mustard, and auburn showcased in the flora, the scene looked like a watercolor dreamscape. And then I finally laid eyes on what some consider the crown jewel of the Mojave Desert: Meandering quietly down the center of the canyon, nourishing the soil that germinates desert flora of all shapes and sizes, was the Amargosa River. I could practically feel how needed the river was, how essential its moisture was to the creatures and plants that have depended on it through the seasons and the centuries. The river, the great mother of the canyon.

For most of its 185-mile length, the Amargosa River actually travels underground, concealed beneath desert washes and the unpaved roads of the American Southwest. When it does surface, it usually appears as a stream, narrow enough for leaping across and shallow enough to keep your shorts dry. Its headwaters emerge near the peak of Black Mountain, north of Beatty, Nevada, and that origin point is an important cultural and spiritual site for the Timbisha, whose medicine women and men approach the mountain on their own kind of pilgrimages.

From Black Mountain, the river runs through Oasis Valley and then skirts south along the Nevada and California border before running underground through several Wilderness Areas. It emerges once again near the modest villages of Shoshone and Tecopa and then runs free into the Kingston Range. This is where nearly thirty miles of it has been given Wild and Scenic River status, which aims to protect the waterway and preserve habitats for endangered species such as the least Bell's vireo and the Amargosa vole. From the Kingston Range, the river heads west toward the Dumont Dunes and then once again buries itself in the desert floor, traveling under Highway 127 and then north into Death Valley, where it finally goes dry under the Badwater Basin.

From the lookout, I carefully scrambled down to the water and then knelt over its shallows, looking for the diminutive Amargosa pupfish and speckled dace, neither of which I could see. I removed my shoes and socks, placed my bare feet in the river, lifted my head toward the sky, and closed my eyes. Surrounded by the parched landscape, where every tree appears malnourished and bushes teeter on the edge of life and death, I struggled to reconcile the sight of the arid desert with the sensation of the cold water beneath me. As the coolness seeped into my pores, I felt an instinctive urge to lie down, soak my dry skin, and drink from the river as if it were the last water source on earth. The sensation reminded me of a few lines written by Edward Abbey about another river in the West: "Come with me, the river said, close your eyes and quiet your limbs and float with me into the wonder and mystery of the canyons."

Eventually, I pulled my numbed feet out of the water and found a nearby rocky ledge to sit on. I checked my hiking app and realized I had walked a little over one mile in three hours, likely my slowest pace since boyhood. I hadn't made it to the waterfall, or to the slot canyon across the river, or to the hanging gardens protruding from mountain springs farther down the trail. Other than the river, the landscape wasn't much different than the one I'd seen in nearby Rainbow Basin. The

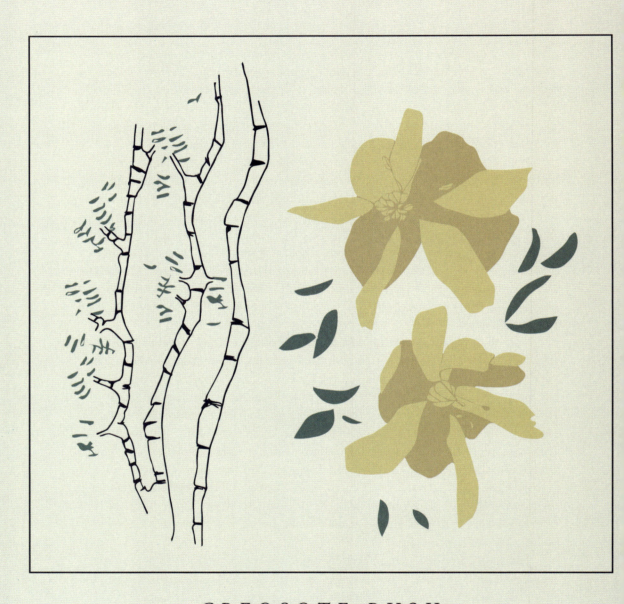

*CREOSOTE BUSH*
Larrea Tridentata

rocks, canyons, hills, and flora in the Amargosa are nearly interchangeable with those of my earlier desert foray a few weeks prior.

But, sitting on the rock in the Amargosa, I felt how different this excursion was. I flipped through several pages of scribbled cursive in my notepad, recalling the findings and sentiments from my slow walk. The birds, the colorful mountain, the coyote, the way it felt to see water in the desert. If I'd explored the Amargosa at my typical hiking pace, taking everything in only at first glance, I might have ended this trip similarly disappointed. But both by deliberately reducing my pace and by studying the Amargosa region beforehand, a more complete history had found its way into each step I took. In noticing, I had witnessed the vibrant biological community of diverse species. In slowing down, I was more attuned to the subtle language of the desert. While I'd return to the Amargosa over the following days, and on subsequent trips, in that moment during my first visit I could already sense a connection growing.

I headed back as the disappearing sun cast its last light on the highest peaks. I pointed my camera to mountaintops turned copper and sandy ridges turned golden and the sullen sky darkening from gray to blue. I arrived to my van sitting alone in an empty parking lot. Darkness completed its takeover as I drove out of the canyon and into the night, looking for a place to camp.

---

After my experience in the Amargosa, I went back to BLM lands in the Mojave and Colorado Deserts on eleven more trips over the course of four winters, the season when visiting these parched locales is less formidable and the risk of dying from heat exposure or getting fanged by a rattlesnake drops to more acceptable levels. I sometimes ventured alone, other times with friends, and I often took my family, introducing my kids to these lands that had come to mean so much to me.

From wrapping up against the cool temperatures at sunrise to shedding layers in the midday heat and then circling back toward camp as the moon started to make an appearance, I have walked almost two hundred very slow miles in those regions. When I set out on this project, I had wondered what kind of scenery I'd find on these BLM landscapes in the desert, and I remember wondering if they even had

Amargosa Canyon

Kingston Range Wilderness

any stories worth telling. It didn't take long for these questions to drop to the wayside in a state of complete irrelevance, fading into the background with each step until they had disappeared altogether. In the accumulating effect of spending long days walking the arid country and cold nights sleeping under the stars, the holiness of the desert began manifesting itself one pilgrimage at a time.

Yes, there was the kind of blockbuster scenery that takes your breath away, the kind that compels you to hug your kids and kiss your lover, the kind that makes you want to scream a loud *thank you* to the universe for such a moment. Like getting to climb inside the 79,000-year-old Amboy Crater. Or running down the colossal, never-ending sand dunes in the Cadiz Dunes Wilderness. And those hundred-mile views! The wildlife! The endemic flora! These awe-inspiring scenes were scattered across lands once deemed unattractive and lifeless.

But there were also more subtle beauties to be found. Like watching the moonrise chase the sunset from one horizon to another, reinvigorating colors on the landscape you hadn't noticed during the day. Or witnessing the delicate red flowers of the ocotillo or smelling the fragrance of creosote following a rain. Or the way the desert feels in the late evening after a long day in the sun, when those cool breezes hum by and the night sky comes alive with stars. There is a collective exhale among all the living. The flora can open their pores, the coyotes can begin their howling, and the humans can head for the porch or the camp chair and rest easy until the sun rises again.

Amidst the extraordinary and nuanced scenery of the desert, I experienced a pilgrimage to the self. Each mile made me confront my preconceived notions of what it means to be immersed in nature. I had always looked for showstopping scenery and pulse-raising adventures around every corner, and I used to feel underwhelmed if I didn't find them. But the desert invited me to leave behind my notions about what I might find or experience, or even how I might feel. A person can go looking for this or for that, whatever it is they have in mind, but the desert doesn't always give you an explosion of grandeur. Holiness here is not a rapture; it is a whisper, reminding you that everything you need to see is around you at every moment.

The great secret of the desert is that it often sings quietly, offering sweet melodies for anyone who will listen.

Sawtooth Canyon Campground

Centennial Canyon

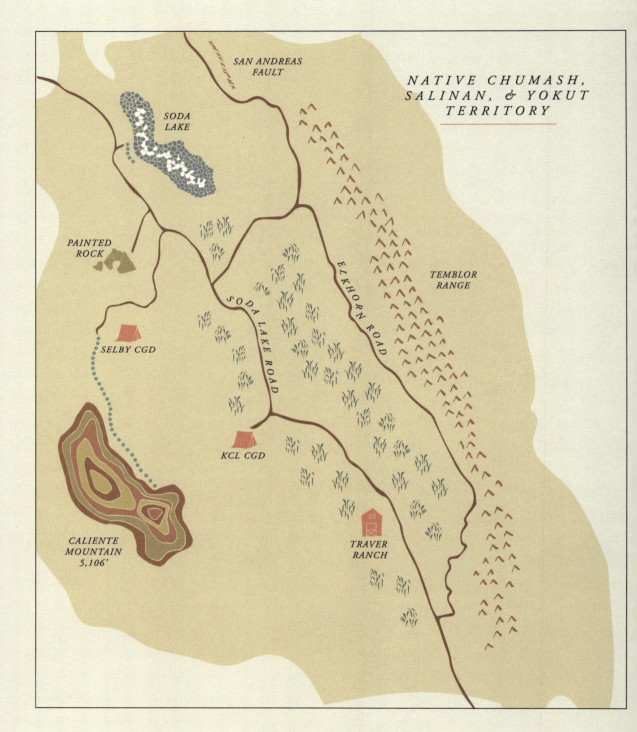

# CARRIZO PLAIN: *Place Attachment*

*It was late fall of 2022.* I was standing near the intersection of Soda Lake and Panorama Roads, at the rough middle juncture of the Carrizo Plain National Monument. The earth was almost perfectly flat, the level terrain broken up only by slight rises and depressions along the valley floor. Only in the distance did the topography start to fold in on itself.

To the northeast were the soft, undulating hills of the Temblor Range, which ran along the horizon like rolling ocean swells. To the southwest was the more formidable Caliente Range, its peaks higher and its slopes steeper. Directly north was the dried-up Soda Lake, calcified and overcooked.

The midday sun illuminated every ridge, canyon, and rock across the plain, the only shadows thrown by distant junipers stretching up the ridgelines. At 59 degrees, the air offered cool respite from what was a brutal October, the weather on this day breaking a lengthy streak of triple-digit temperatures. A steady wind at fourteen miles per hour swept across the landscape.

Everything was covered in a thick layer of dust, including me. The chalky, bitter taste of earth hit my tongue and tiny particles of grime lodged in my pores. With enough time, I thought to myself, the landscape could envelop me like a cocoon and bring me back as a saltbush or lupine.

The dominant feeling was one of isolation, something I have to come to expect from my time on BLM lands. There were no vehicles bouncing across the rugged roads, no humans walking the trails, no cattle grazing the hillsides. Even the herds of tule elk and pronghorn and the hermitic coyotes that roam the plain seemed to have retreated to dens and canyons and higher elevations, out of plain sight, burnt out from the long summer.

It had been two hundred twenty days since the last rain fell, and the yearly total to date was a staggeringly low 1.59 inches. Severe drought was closing in on four years. The entire valley and surrounding hills took on the color of a hay bale.

But despite the barrenness before me, I knew that this land had a sleeping magic, the type hidden like seeds in the soil.

    Here's another detail I couldn't see along the cloudless horizon: Somewhere over the Pacific Ocean, a storm was slowly moving east toward the Carrizo. In a few days' time, nearly an inch of rain would fall, an autumn rarity. It would be the first in a series of atmospheric rivers that would bring a whopping thirteen inches of rain over the next four months, giving those parched seeds one crucial ingredient of the divine potion they needed to make their way up to the sun.

Soda Lake

CARRIZO PLAIN: PLACE ATTACHMENT   59

The 204,000-acre Carrizo Plain National Monument was designated in 2001, following years of collaborative efforts between the BLM, the Nature Conservancy, and California Fish and Game (now the California Department of Fish and Wildlife). Despite its relatively small size, the valley of Carrizo is the largest remaining area of grassland in California, making its protection against potential development crucial. Until the 1980s, much of the Monument was owned by Oppenheimer Industries, which leased land to local ranchers for grazing. However, the arid conditions and

lack of reliable water supply made ranching economically challenging, and as job opportunities emerged in other regions, the ranching community dwindled. By the time the Monument was designated, after the BLM and the Nature Conservancy had purchased the land from Oppenheimer, only a half dozen ranchers remained on the plain.

Today, the greatest concern in the Carrizo are the eighteen oil and gas claims covering nearly half of the Monument. These claims, established before the Monument's designation, are still valid. Throughout most of the twentieth century, developers digging in search of black gold found a few active wells in the southern reaches of the Carrizo, but otherwise their efforts mostly ended in dry holes. However, with rising oil prices and advancing technology, companies are once again exploring for oil, leaving existing protections of the grasslands in a precarious state.

The Carrizo is roughly fifty miles long by fifteen miles wide, a rectangular body of land perfectly framed by two mountain ranges running on either side of the plain. Thanks to this uniform arrangement and clear access points, getting your bearings is relatively easy. There are just two main entrances to the Monument: one from the north and the other from the south, both connected by Soda Lake Road, a dependable but bumpy gravel thoroughfare that runs straight through the long valley. While narrower routes crisscross the plain, Soda Lake Road does most of the heavy lifting, providing access to several main attractions and the Goodwin Education Center, which offers information, brochures, and exhibits.

The Carrizo provides as close a picture as one can get to what the Central Valley looked like when the land was tended solely by the Yokuts, Chumash, and Salinan tribes. Despite the unfathomable population losses suffered in the nineteenth and twentieth centuries due to disease, famine, and reservation life, Indigenous traditions and cultural practices continue to this day, including at a special sandstone outcrop called Painted Rock.

While it's only one of dozens of outcrops along the Caliente foothills, this particular formation houses some of the most elaborate Native rock art found in North America. From above, the outcrop takes on the shape of a horseshoe, with high walls protruding from the earth like a fortress protecting the sacred panels within. The only entrance is from the north, and entering it feels like stepping through a portal into another realm. Inside, the smooth rock faces are covered in holes perfectly sized for nesting barn owls. Orange and green lichen are erratically smeared

Caliente foothills

across the boulders like misplaced brushstrokes. However, the most striking feature are the colorful murals, featuring figures and symbols painted over millennia by the Indigenous peoples who have called the grasslands home.

The painted colors contained in the murals include black, red, and white, made from local minerals such as hematite, shale, and gypsum mixed with animal fat or oil extracted from milkweed, seeds, and bird eggs. Many Native people believe the art was painted by shamans, who saw the rocks and caves as portals to the sacred realm, where spirits could be influenced to stay away from or intervene in human affairs.

In 1982, Leslie Schupp Wessel, then an anthropology student at Cal State Northridge, completed her master's thesis on the pictographs at Painted Rock, summarizing her extensive research with this poetic conclusion: "The paintings on old rocks in California speak of an alien world, of a sun and moon now de-mystified, of bears which no longer roam the hills. Only the coyote and rattlesnake remain, to howl and hiss a warning that the earth has other life besides man."

*TEMBLOR RANGE*
Carrizo Plain

These warnings about "other life" are just as relevant today. The Carrizo Plain provides a vital refuge for an unusually high concentration of endangered species, including the San Joaquin kit fox, the giant kangaroo rat, the blunt-nosed leopard lizard, and the California jewelflower, among others. Since the mid-nineteenth century, the demands of our growing population have paved and plowed over most of the Central Valley, fragmenting habitats and leaving these species with nowhere else to go.

---

While the protection of natural features, Native cultural artifacts, and imperiled species was the main reason behind efforts to have this area designated a National Monument, I was drawn back to the Carrizo for something else: the seeds. When weather conditions align, typically only once or twice a decade, millions of dormant seeds burst forth in such vibrancy that the golden-brown grasslands of autumn transform into a display of wildflowers so vast that distant satellites can detect the colors from space. Botanists once referred to the springtime event as the March Miracle, but today it is mostly known as the superbloom.

Five months after the first atmospheric river hit in November 2022, and one month after the last rain fell, the empty and desolate plain I had experienced in the fall had thoroughly and supernaturally transformed. The landscape was so incomprehensibly different I couldn't help but stare in disbelief. Soda Lake Road had become a yellow brick road, and I had wandered into the Land of Oz. I couldn't figure out where to look, as if my eyes and brain had gone haywire into a hyper state of mesmerizing distraction that left me utterly speechless. It was like taking Dorothy's first step into the technicolor world of Munchkinland after living in sepia-toned Kansas.

Thanks to a field of bright-orange California poppies, there were already cars lined up at the Monument's entrance. Happy couples stood on the outskirts of colorful fields and took selfies, flashing smiles and peace signs. Kids wandered in just about every direction, running this way and that, wild and delirious after long car rides, and looking as thrilled by the sea of open space as they were about the flowers. I knew the bloom would attract a multitude of people, but I was still stunned by the presence of so many families on BLM land. Two and a half years into my

Painted Rock

project—which had already taken me on nineteen trips across the Golden State—I had never seen anything quite like it. I slowed down as I drove past, waving and smiling at the giddy onlookers.

Past the entrance, I was greeted by vibrant fields of Great Valley phacelia, congregated so densely that the herbaceous flowers seemed to blend together into a single expanse of dark purple. Surrounding the phacelia was an unending sea of orange and green brought on by an outbreak of fiddleneck, arguably the second-most-dominant wildflower to proliferate on the plain. As a vast assembly, the fiddleneck appeared somewhat uniform and ordinary, but closer inspection revealed individual plants of wild eccentricity, with light-green stems rising from the earth in every direction but straight, pointing this way and that, fighting for sun, reaching differing heights, and with curling leaves and tiny hairs growing along their robust branches. At the top of the stems were coiled buds, and above those were the five-star petals of the fiddleneck flowers. The orange and yellow blooms sprouted forth from stems curled into the shape of a seahorse. These "land horses," as I affectionately called them, filled in vast sections of the valley floor.

The rest of the plain could only be described as an exuberant medley of yellow, as if that winter's atmospheric rivers had rained down droplets of dye, coloring the Carrizo in lemon, amber, mustard, and butterscotch. The flamboyant display was thanks to several high-achieving flowers, namely the goldfield, the woolly sunflower, and the tidytip.

It was late morning by the time I arrived at Selby Camp, my home for the weekend. After moseying north on Soda Lake Road for thirty miles, transfixed by the bloom, my mood was one of awe and reverence. Campers of every stripe filled the campground, a sight that left me hopeful. I met mountain bikers, bird watchers, wildlife photographers, botanists, and hikers—a small but diverse group of nature enthusiasts. One older gentleman had me particularly transfixed, his mood a perfect symbol for the reverence I felt. He was sitting by himself in a metal lawn chair outside his dilapidated trailer, his body pointed away from the campground and into the hills. No phone. No book. No binoculars. He was simply looking.

There are some who loathe the crowds and wish for all the beautiful places to remain hidden, hoping to keep their favorite landscapes all to themselves. While I understand this sentiment, especially when more people can mean more trash and disruption, the reality is that the balance between gatekeeping and conservation is delicate.

When oil prices spiked in 2007, an oil and gas corporation called Vintage Production sought to explore their oil claims using invasive "thumping machines" that would have severely disrupted kit fox and kangaroo rat habitats. In response, a groundswell of public support pressured the BLM to require stricter environmental assessments, ultimately preventing the thumping from taking place. Similarly, in April 2017, when President Trump ordered an unprecedented review of twenty-seven National Monuments, including the Carrizo Plain—and at a time that coincided with the 2017 superbloom—there was a public outcry for continued protection of the Monument. In both cases, threats were averted thanks mostly to people who had experienced the Carrizo firsthand. That said, if isolation is what you seek, the Carrizo is almost empty in the months and years outside of the bloom.

I was back in the Carrizo both to enjoy the exuberance of springtime flowers and to learn about the conditions that lead to a superbloom. I figured all the rain was a big part of it, but I assumed there had to be more to the story. Even in other years laden with moisture, the blooms hadn't been nearly as abundant. How on earth had the dusty, blank canvas I had experienced just a few months earlier turned into this eruption of color?

To find the answers, I set up an afternoon walk with Emma Fryer, a botanist from nearby San Luis Obispo who had made the wildflowers of the Carrizo the focus of her master's thesis at Cal Poly. Part of this long-term research included studying the wildflowers, collecting their seeds, and conducting elaborate greenhouse experiments. I was lucky to catch her before she moved to Switzerland, where she is embarking on her PhD at that country's Federal Institute of Technology (ETH Zurich). I was drawn to her expertise not only because she had a breadth of experience in the Carrizo but also because she draws beautifully detailed flower portraits. She is as skilled an artist as she is accomplished at botany.

We parked along Simmler Road, on the east side of the ephemeral Soda Lake,

SAN JOAQUIN KIT FOX
*Vulpes Macrotis Mutica*

which was at that time filled to the brim thanks to the deluge of winter and spring precipitation. From the surrounding hills, the expansive lake took on the appearance of a river delta, with its many inlets and tributaries extending into the surrounding saltbush and shrubs. I grabbed my notepad and camera and stuffed my pockets with a handful of tissues, the only weapon I had to fight the pollen-induced allergy attack that was building with each hour I spent on the plain. Emma wore black jeans, a green T-shirt, and a faded hat that appeared to be slowly deteriorating from years of exposure to sun, blowing dust, and sweat–an appropriate symbol for a field botanist. Around her neck hung a mini magnifier called a loupe or a hand lens, the primary tool used for identifying flora species.

As we set off on our delicate walk around the flowers, she radiated joy and a healthy dose of surprise, which caught me off guard. She reminded me that, until this spring, she had only witnessed the plain in an arrested state of decay.

Temblor Range

Despite having visited the Carrizo many times before, this abundance was as new to her as it was to me.

We walked slowly along the road and followed narrow pathways that led in all directions, some dead-ending at the lakeshore and others simply running into a wall of flowers. After spending the past few years approaching BLM lands with the mindset of a pilgrim, I was now accustomed to the slow, easy pace of noticing, and as we wandered, Emma described the soil and floristic features in vivid detail, happily translating the scientific plant names for an amateur like me. I learned about erodium patches, lichen-hosting "slickspots," and smectite clay. She showed me the subtle differences in petals between a goldfield and a wooly sunflower, which look identical but for the slight variances in color and petal shape.

Her thorough knowledge of Carrizo botany and her love for the native species thriving there were infectious. When I asked her why studying flowers was her

*MENZIES'S FIDDLENECK*
*Amsinckia Menziesii*

special focus, as opposed to trees or shrubs, she noted with tenderness that she "likes to see the whole thing." The bliss in studying flowers, she told me, is the ability to hold and examine the plant in its entirety, from root to tip.

But it was her explanation of the superblooms' great mystery that finally unlocked my understanding. She explained that, prior to the bloom, millions of dormant wildflower seeds lie below the surface, in parched soil, simply waiting for the right conditions to germinate. These little miracles are physiologically resilient and can remain in deep dormancy for decades—a time frame I found astonishing. Unlike the invasive species that often crowd them out but lack the biological adaptability to survive prolonged droughts, these native seeds endure. As the dry years add up, the invasive seeds die off, paving the way for what we were witnessing. When weather conditions are just right, the resilient seeds of the poppy, bluebell, fiddleneck, lupine, phacelia, owl's clover, California jewelflower, desert candle, and goldfield awaken from their hibernation, setting off the explosion of color before us.

A miracle if there ever was one.

By the time our long afternoon of wandering had come to a close, Emma's black pants had brushed up against enough goldfield and fiddleneck that they now looked partially dyed in yellow and orange. As I packed up my camera gear, I watched her stare out at the plain with longing in her eyes, perhaps thinking of her impending move. "I'm going to miss this," she said quietly. After repeated visits and years of studying the flowers of the Carrizo, the landscape had fully endeared itself to her.

---

In environmental psychology circles, this emotional bond between person and place is known as "place attachment." The theory suggests that people who develop significant relationships with an environment experience a greater sense of identity, belonging, and well-being. Attachment can manifest in various forms, from connections with the physical dwellings where we eat, sleep, and socialize, to affinities for the neighborhoods and towns we call home. For many of us, like Emma in the Carrizo, an essential aspect of place attachment is our connection to nature. Sometimes the nature we connect with spans whole regions, like the Mojave Desert or the Eastern Sierra. Other times, these connections are with National

Soda Lake

Simmler Road

and State Parks or, at a smaller scale, with specific trees or waterways, even those in non-wilderness environments.

Often, this relationship with nature develops over an extended time, especially as repeated visits allow us to form deeper familiarity with a landscape or feature. Unique experiences can also play a role; reaching a mountain summit, stargazing, catching a first rainbow trout, and encountering wildlife are moments that can create attachment. Even challenges like facing inclement weather during a camping trip or enduring a strenuous hike can strengthen our bonds with nature. And, as I learned through my pilgrimages in the desert, curiosity and knowledge are essential to forming strong connections. These bonds help transform us from mere tourists into individuals with a more immersive attachment to the natural world.

I have thought a great deal about place attachment while exploring BLM-managed environments. Almost all of these public lands were new to me, locations that used to be just names on a colorful wall map hanging in my living room. But now, after nineteen trips, I have formed a connection with them. In the Carrizo, the attachment came not from one trip or event but through a collection of moments, days, and seasons, all stitched together. I was obviously impacted by the unrivaled display of wildflowers in spring, but an equal part of my love for the Carrizo came out of having encountered the landscape during an autumn drought, when the valley appeared nearly lifeless, its surprises hidden away.

The psychologist Edward Relph, in his 2022 book *Place and Placelessness*, expands on this concept by calling place attachment a "fundamental human need." This raises an intriguing question: Is our interaction with nature as vital as our basic needs for shelter, safety, and sustenance? It certainly feels that way. However we find ourselves connected to a natural environment, and wherever that place may be, this transformational love affair between people and nature can be vital to our wellness. It grounds us, provides solace, and offers a space in which we can take a long, deep breath—the great antidote to the pressures of life. And this bond can also be integral for the very places with which we have these connections. Just as the Carrizo had people to stand up against threats to the area's species and grasslands, so many of these lesser-known BLM landscapes need more humans who have experienced and formed bonds with them. Without such attuned advocacy, many extraction and development projects go forward unchecked.

If the holiness of the Mojave laid bare my expectations for both nature and my experiences in it, multiple trips to the Carrizo had pushed me to dig deeper, to sit in

Silver lupine

Tidytips

the grassland and consider how my burgeoning attachment with these places might alter my relationship to them. Learning to see and appreciate the subtle beauty of these BLM lands was transformational, and part of that transformation was the uncomfortable realization that my relationship with nature had, until then, been largely one-sided: a take-take-take kind of union.

---

As Emma packed up for her long drive home and I searched for more Kleenex, I thanked her for the illuminating walk. Back at camp, I decided to take a sunset hike in the hills rising up from the plain. I walked along forgotten fences, rocky outcrops, and dried-up creekbeds. My only company was a yellow-bellied western meadowlark who followed me from fence post to fence post, serenading me along the way. When I got back to camp, the older gentleman was still sitting in his metal lawn chair, still looking out over the hills. Taking his cue, I decided to plop down into my own chair and stare out over the scene in silence, watching the last of the fading light. The junipers cast long shadows, and the earth itself seemed to change colors like a flip book—from gold to orange to forest green—before giving way to the cool midnight blue of darkness.

Caliente foothills

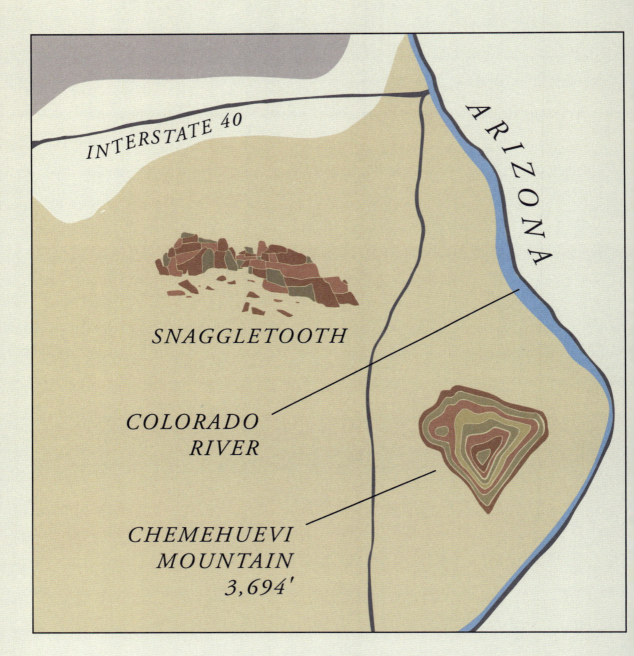

# BORDERLANDS: *The Story of Hunting*

*"I'm in the middle of a meat crisis!"* Aaron yelled from his doorway with a voice full of both urgency and excitement.

"Meat crisis?" I inquired, seeking clarification as I stepped out of my van.

"My freezer is almost empty, and I hate buying meat at the store. It's just not the same."

Aaron's enthusiasm was the reason I found myself standing in his Orange County driveway. I was about to join him for a three-day deer hunting trip back to BLM lands in the Mojave. Thanks to our day jobs occasionally intersecting, I had spent numerous lunches with Aaron and his business partner, Ryan, over the years. By day, Aaron Shintaku and Ryan Haack are seasoned photographers, but they also have a hobby that never fails to supply our meals with captivating stories: They are both avid hunters.

As I sought to understand the range of allies working to protect BLM lands, I became increasingly aware of the pivotal role hunters and anglers have long played in land and wildlife conservation. This powerful coalition of recreationists has sometimes been overlooked, misunderstood, or ignored in environmental circles. From an outside perspective, I had wondered if most hunters disregarded issues of biodiversity, sustainability, and ecosystem health in their pursuit of game, but as I delved deeper and learned about their contributions to conservation ideas I cared about and to landscapes I had experienced firsthand, my preconceptions were challenged. I realized that hunters and anglers had their own forms of place attachment to nature and wildlife, and I was keen on broadening my perspective. I figured the best way to better understand their connection was by immersing myself in an actual hunt.

While Aaron and Ryan usually hunt deer on public lands managed by the US Forest Service, they agreed to adapt their annual trip to include a hunt on BLM lands, graciously allowing a newbie like me to tag along as an observer. We set aside three days for the trip, although Aaron reminded me, "We actually have two sunsets and two sunrises." These are optimal times for witnessing the elusive desert mule deer, who are mostly on the move during first and last light. When talking about this species, *elusive* is a key word; Aaron and Ryan warned me that the odds of bagging a deer in the desert are much lower than in other landscapes.

The California Department of Fish and Wildlife manages hunting across the state, categorizing all public and private land into specific hunting zones. Each zone is allocated a certain number of "tags" per animal species, and that number determines how many animals can be harvested in that area. The department uses a variety of methods, metrics, scientific principles, and expert guidance to inform these allocations. In the D12 zone, for example, where we would be hunting, a total of 950 mule deer tags are released every year. In 2022, only 57 deer were killed, or 6 percent of the maximum.

Determining the appropriate number of tags to release each year is an exercise in patience and adaptability. Too many tags can negatively alter animal populations, and too few tags can mean certain populations increase to the detriment of other species of flora and fauna. It's a tricky game of numbers that is always subject to annual and decadal adjustments to ensure the health and sustainability of wildlife populations and their habitats.

Under a clear November sky, we loaded up my work van with hunting gear and equipment, which included a heavy-duty backpack brimming with knives and tools, a large cooler prepped with slabs of ice, a foam archery block for target practice, and two large bins filled with enough camouflage attire to hunt anywhere from the Mojave Desert to a blizzard in Alaska. For safety, Aaron had me order a bright-orange vest, which matched the color of his hat. Donning orange is a critical safety measure for hunters and other recreationists, ensuring visibility and preventing accidents in shared landscapes.

But the real prize was the compound bow. Having never seen one up close, I was stunned by its weight and mechanisms. This was a precisely engineered machine made of carbon fiber, aluminum, pulleys, and cables, and the carbon-tipped arrows can travel at roughly three hundred feet per second, enough force to easily take down a deer (or even a bear). With the van fully loaded and our D12 tags secured, we hit the road.

The long history of hunting in California is marked by a series of peaks and valleys. Over the last two centuries, the narrative unfolds in three distinct acts: an era of symbiosis, a period of exploitation, and a chapter of redemption.

For the earliest Native inhabitants and many of their living descendants, hunting has been a deeply spiritual practice. For these communities, animals are not merely resources but sacred beings, integral to the very fabric of existence. This reverence has inspired a form of hunting grounded in respect, gratitude, and stewardship, ensuring the sustainability of ecosystems for generations to come. Such practices are part of a broader system of Traditional Ecological Knowledge, which has for thousands of years maintained biodiversity and ecological balance through careful management of the landscape and its animals.

In the nineteenth century, these Native values were challenged by American settlers driven west by the doctrine of Manifest Destiny, an ideology predicated on the concept that expansion across the continent was divinely sanctioned by the Christian God. These settlers viewed the land, plants, forests, and animals as entities to be conquered and subjugated. In sharp contrast with the Indigenous practices, the settlers' approach was marked by a voracious appetite for market hunting, sport hunting, and land clearance. Their overhunting, coupled with habitat destruction, triggered a severe decline in wildlife populations, as epitomized in California by the extinction of the grizzly bear and the near extinction of endemic species including the tule elk and some species of beavers, wolves, and egrets.

At the dawn of the twentieth century, in the devastating aftermath of this western expansion, the conservation movement began to take root. Conservation advocates—from grassroots enthusiasts such as John Muir to high-profile figures including President Theodore Roosevelt—popularized the preservation of natural spaces, the establishment of parks and wildlife refuges, and the enactment of groundbreaking legislation, which included the Lacey Act of 1900 and the Migratory Bird Treaty Act of 1918. These acts marked the start of federal regulations aimed at shifting hunting practices away from market hunting and toward promoting sustainable management of wildlife populations for sport hunters.

Building on this conservation mindset, hunters and anglers initiated a radical plan: funding wildlife preservation by taxing themselves. Through the Duck Stamp Act (1934), the Pittman-Robertson Act (1937), and the Dingell-Johnson Act (1950), hunters and anglers pay a 10 to 11 percent excise tax on firearms, ammunition,

Turtle Mountains Wilderness

archery equipment, fishing gear, and boats, an unprecedented funding model that directly connects the people who use the land with the conservation of that land. Collectively, these laws have raised billions of dollars for wildlife recovery, habitat restoration, and educational programs.

Aldo Leopold took the conservation concept even further. Celebrated as both a hunter and a naturalist, Leopold, in his revolutionary book *A Sand County Almanac* (1949), called for a holistic approach to conservation, urging people to accept the call as a moral responsibility. "The land ethic," he wrote, "simply enlarges the boundaries of the community to include soils, waters, plants, and animals, or collectively: the land. . . . In short, a land ethic changes the role of *Homo sapiens* from conqueror of the land-community to plain member and citizen of it." Leopold's land ethic was, in many ways, a return to the ancient wisdom of the land's original custodians.

In addition to advocating for a new land ethic, Leopold addressed some of the consequences that come when certain styles of hunting and wildlife management prioritize game species such as deer and elk, sometimes at the expense of carnivores including wolves, bears, and mountain lions. This imbalance can disrupt ecosystems, affecting predator-prey relationships and non-target species. Leopold's humility, his ability to look inward and allow his perspective to evolve, is noteworthy. Reflecting on the impact of allowing hunters to kill wolves without limit, he wrote, "I thought that because fewer wolves meant more deer, that no wolves would mean hunters' paradise." After observing the devastating effects of unregulated hunting—such as a significant increase in the deer population, which led to overbrowsing of vegetation and eventually to widespread starvation among the deer—he concluded, "I now suspect that just as a deer herd lives in mortal fear of its wolves, so does a mountain live in mortal fear of its deer."

Conservation efforts today continue through an effective coalition of organizations committed to responsible stewardship. Leading the charge are nonprofits such as the Theodore Roosevelt Conservation Partnership, the Rocky Mountain Elk Foundation, and Backcountry Hunters and Anglers. These groups provide opportunities for hunters and anglers to deepen their connection with the landscapes through volunteer events, mentorship programs, work on policy initiatives, and educational workshops. These groups are also crucial allies in advocating for the protection of BLM lands, and their efforts help to ensure continued access, healthier wildlife habitats, and cleaner waterways, all of which benefit people who hunt and fish on public lands.

Against the backdrop of learning this rich conservation history, my path intersected with Mark Kenyon, a prominent figure in the hunter-conservationist movement. In his book *That Wild Country* (2019), Kenyon seamlessly interweaves his personal narrative as a hunter with a wealth of grounded insights and eloquent discourse on the numerous challenges facing public lands. I hoped talking with him would further my own understanding of the relationship modern hunters have with conservation efforts.

When I asked about this intersection, Kenyon articulated a clear and undeniable bond between the two. For him, there is no taking from the land without giving back to the land. "If we're going to hunt and fish, if we're going to insert ourselves into that circle of life," he told me, "it's going to be a really serious thing. It's not trivial, not a video game, not something that we can take lightly. If we're going to do this, we're going to do it with care and respect. We're going to do it ethically. And we're going to do it with very clear eyes about what we're doing. We are trying to take an animal's life. And this meat will feed our family. And then, in the end, here's how we're going to make sure that this can continue to be done in the future, how we can help steward the wildlife and protect these places."

For Kenyon, who is helping bridge the legacy gap between Roosevelt's and Leopold's times and the pressing needs of today's challenges, like biodiversity loss, hunting is a circular pursuit. While hunters contribute monetarily through self-taxation and by purchasing tags and permits, Kenyon believes that deeper change happens through the reciprocal act of volunteering. "I think the thing that leads to more transformative change is to actually get out there and get your hands dirty," he said. Through his popular podcast, *Wired to Hunt*, he not only shares deer hunting tips and strategies but also uses his platform to encourage listeners to find their own ways to volunteer, including through the aforementioned organizations, who are making it easier than ever to interact with the land.

For me, the most impactful part of our conversation came as Kenyon shared his philosophy on hunting, which he sees as a conduit to a deeper, more personal connection between people and landscape. "Hunting changes your relationship with the natural world and with wildlife," he said. "You go from being this person that passes through to now being a part of that ecosystem. You're not just an observer,

88 THE ENDURING WILD

North Algodones

Willow Creek Canyon

you're actually a cog in the wheel, like the wolf or the fox or anything else out there. Now you're taking from the land and inserting yourself into the circle of life. And when you're actually participating in life and death out there, it completely changes the color of the experience."

As someone who has primarily experienced nature as a hiker, the idea of *inserting* myself into the circle of life, rather than simply *observing* it, marked an important shift in my understanding of the hunter's mindset. My many pilgrimages through the Mojave had taught me to slow down and notice the world around me, but now I was headed back as a hunter instead of a hiker. I wondered how this shift might alter the color of my own experience in nature.

---

Four and a half hours after leaving Orange County, we arrived at the 85,840-acre BLM-managed Chemehuevi Mountains Wilderness, just south of Needles, California. The Colorado River runs along the eastern edge of the Wilderness, dividing the arid borderlands of California, Nevada, and Arizona. We chose this area because I had witnessed a herd of deer there on a previous trip and because the Wilderness allowed us to hunt without the distraction of noisy roads, which aren't permitted in federally designated Wilderness Areas.

It was the week before Thanksgiving, but the weather had only recently caught up with the autumn season. Two weeks prior, temperatures were still hovering dangerously close to three digits, in what proved to be the final gasp of summer. Now, highs were in the 70s, the kind of respite that desert residents everywhere celebrate after half a year of stifling heat.

While Aaron gave me a quick hunting lesson, which basically amounted to keeping quiet and following his lead, I watched as he methodically prepared for the pursuit. He was calculated and thorough, and suddenly more serious than I had ever seen him before. He donned camouflage attire from head to toe, including a heavy vest with pockets that carried a range finder, a walkie-talkie, a wind indicator, and binoculars. His rugged backpack, also in camouflage, contained the tools, knives, and straps we would need in case we killed a deer. It also secured his hunting license and a D12 tag that allowed him to bow hunt in the zone. The number and weight of Aaron's accessories was substantial compared to what I was carrying, which included binoculars, a small day pack with water, and a hip pack containing a wilderness whistle, a phone, a notepad, and a pencil.

I was fascinated by Aaron's meticulously organized compartments and gadgets, each item carefully stowed in its place. Naively, I asked, "Where do you put the bow?" Aaron's eyes flashed with a look of incredulity before softening into patience. "In my hand," he explained with a smile. "Sometimes all you have is a second or two. You gotta be ready."

He then went over our objectives for the afternoon, laying out a plan that went beyond just the pursuit of a deer. Our aim was to acquaint ourselves with the terrain and find a vantage point from which we could survey deer emerging from their daytime beds. Then he asked me the question that precedes every leap into the unknown: "You ready?"

The question hung in the air, a challenge and an invitation all at once. Suddenly, the moment I had been thinking about for months was right in front of me. As we took our first steps away from the van, I found myself in a completely different "outdoors mode" than I had ever been in. I was now used to setting off as a pilgrim, walking and noticing, observing plants and rocks, and hoping to catch a glimpse of wildlife from a distance. But Aaron's bow, hanging from his clenched fist, reminded me that I was no longer just a casual observer. Kenyon's words acted as a solemn reminder: "We are trying to take an animal's life."

The desert mule deer, or "muley," as it is often called by hunters, is as exquisite looking as its black-tailed cousins that roam closer to the Pacific Ocean. Adult males are tall and broad-shouldered, weighing as much as 330 pounds. Their fur is a blend of eggshell white and sandy brown, perfectly camouflaged for the desert. I couldn't help but imagine seeing one dead and wondered what the fur might feel like. I took a deep breath and jotted excitedly in my notepad, "Here we go!"

After a quarter mile, the intensity had already worked its way down from my mind to my chest, where my heart pounded. A nervous excitement spread through my body. My senses sharpened, with subtle changes in smell triggering alerts in my brain, while my eyes scanned the landscape, tracking even the smallest movement. I scolded myself for each step that made too much noise. Despite having walked hundreds of miles across the desert, I had never felt so attuned to a landscape.

Chemehuevi Mountains Wilderness

Chemehuevi Mountains Wilderness

Stretched out before us, in a gentle upward slope, was a canvas painted with everything you might expect from a southwestern desert. The Chemehuevi Mountains have striking white granite peaks in the west that give way to dark-red-and-gray volcanic spires and mesas as the range moves east toward the Colorado River. The afternoon sun beamed between streaks of wispy cirrostratus clouds and a few heavy cumulonimbus clouds floating through the sky. The shadows and sunlight alternated, as if the heavens were flicking a switch off and on.

Flourishing in thick hillside patches were the fearsome teddy-bear chollas, which, with just the faintest hint of sunlight, glowed like Christmas trees. From a distance, the millions of delicate thorns protruding from their stems appeared as soft and inviting as my kids' stuffed animals. But these silvery-white spikes are actually as menacing as a momma bear protecting her cubs. This species is often referred to as the "jumping cholla," for its needles' ability to dislodge from the host branch and attach themselves to the arms and legs of hapless travelers. Once the needles make contact with skin, each microscopic barb hooks in with a hold so strong it often requires the victim to remove them with tweezers. I had already been stabbed too many times to count and avoided the teddy-bear cholla at all costs.

During a short break to scan our surroundings, I whispered a question I had forgotten to ask Aaron earlier: "What happens if we see a buck?"

"That's just the beginning," he warned in hushed tones. "If we are lucky and he hasn't seen or smelled us first, we'll try to find a place for you to track his movements while I attempt to sneak up on him. You can radio me locations and updates. Sometimes the whole process can take hours just to get within shooting distance. It's a game of patience. And even after all that, I still might not have a clear shot."

After a mile of quiet steps across the undulating rises of the desert floor, Aaron pointed to a hilltop in the distance, and I followed him up until we arrived at a few crimson outcrops of coarse rock. We crouched down, and Aaron whispered instructions about how I should be using my binoculars. "You're watching for any type of movement, anything that seems out of place. If the deer are moving, this will be the time."

With an hour until sunset, we perched ourselves on opposite sides of the rock and started scanning. From our new vantage point, the long views over the Mojave were simply stunning. Rounded hills and sharp peaks rose up in every direction, layers upon layers stacked on top of each other, each one silhouetting the next. In

BORDERLANDS: THE STORY OF HUNTING    97

Entering the Wilderness

another departure from my previous trips, I had to remind myself I wasn't just there for the scenery.

As I sat in silence, I became obsessed with the job I had been given, scanning the terrain by mapping it into an imagined grid of lines and squares, like city blocks. I used my binoculars to methodically "glass" up and down the territory, scrutinizing every wash, outcrop, and creosote bush. Despite not sighting even a single living creature, I was surprised by my hopeful confidence. Every moment presented an opportunity, and I glassed with abandon.

As the emerging darkness eventually rendered the binoculars useless, the sky put on a splendid show of color. The ominous, pillowy clouds turned red and orange until the sun disappeared and gave way to the kind of ecstatic magic hour I had fallen in love with during my pilgrimages to the Mojave. Stripes of pink and purple ran across the horizon, above which ran a nearly imperceptible gradient of rose, salmon, plum, and indigo from earth to sky.

"Not a bad way to end the first day," said Aaron, who was instantly more relaxed now that our hunt was on pause until morning. As we descended the slope, we turned our attention to the main event of the evening: dinner.

---

We walked back to the van by headlamp and then drove a few miles west until we arrived at Snaggletooth Rock, where there are several BLM-managed dispersed campsites along the hard rock of the desert floor. We met up with Ryan, who had been scouting farther south, and traded stories while setting up camp and building a fire. Ryan, who was not in a meat crisis thanks to a recent bull elk kill in Idaho, brought out three elk steaks from a cooler and laid them across a metal griddle that sat on the fire.

While the steaks sizzled, Ryan offered a riveting play-by-play of the elk hunt that had provided the dinner we were about to consume. It had happened on the last evening of a six-day hunting expedition on National Forest land in southern Idaho. After hiking dozens of miles up and down mountains, through a landscape of aspen, sagebrush, and pine, he had yet to get close enough for a shot. Bow hunting, he explained, required a level of intimacy not associated with gun hunting.

TEDDY-BEAR CHOLLA
*Cylindropuntia Bigelovii*

Snaggletooth Rock

With a rifle, anything within a thousand feet offered a realistic opportunity for a shot; with a bow, that distance shrank to a few hundred feet. And getting that close to an elk—especially when you factor in the animal's senses of sight, sound, and smell—is an art form that demanded a seasoned portion of teamwork, ability, and instinct.

I listened intently as he walked me through the moment he and his hunting partner entered a pocket of timber and then found themselves staring at an elk herd only forty-five yards away. In a matter of seconds, Ryan raised his bow and let an arrow fly. It was a direct hit to the lungs. The bull took off through thick brush but only made it fifty feet before death caught up to him.

As the tale unfolded, I scrolled through images from the hunt on his phone. One particular picture had caught my eye. It was of Ryan kneeling beside the elk, holding it by the antlers and looking happily exhausted. I've seen versions of this picture dozens of times—a hunter with their kill—but only then did I realize that the image captured just a small part of the experience.

I couldn't help but be struck by the journey it took for an animal in Idaho to end up as a steak on our fire in the Mojave—not just in the geography sense but also in the sense of the hunters' dedication and sacrifice. At home, acquiring beef steak requires a five-minute trip to the grocery store and a ten-dollar bill; the mental and emotional distance between the cow and me is a yawning gulf. But these sizzling steaks were something altogether different: the culmination of extensive time, thousands of dollars for permits and trip expenses, and a significant amount of sweat equity. The relationship between Ryan and this elk was built on a deep respect for the animal and gratitude for the long supply of healthy nourishment it would provide.

Together, we raised a glass for the elk, who not only provided us dinner that night but would also feed a family of six (plus some friends and neighbors) for half a year. It was the best steak I have ever tasted.

As the fire died down, we made plans for the following day. Ryan was on board for a sunrise scouting trip but had to depart early to pick his kids up from school. Aaron and I decided that we would go on to explore the primary wash that cut deep into the mountain's heart, lured by the promise of a natural spring I had located on my offline maps. The journey there and back would require a near half marathon of hiking, but the hunting potential associated with the spring was too enticing to pass up. We set our alarms for 4:30 a.m.

### DESERT TARANTULA
*Aphonopelma Chalcodes*

Overhead, skies were clear. Despite the sun hanging unimpeded, the temperature had trouble hitting 70 degrees, a welcomed gift. The air carried the musky smell of creosote, which was delivering its special scent due to some light overnight rain. A nearly imperceptible wind blew gently in our direction. "Perfect conditions," said Aaron. "They won't be able to smell us coming."

We trudged up the sandy wash in silence, stopping every few minutes to climb a short hill for extended viewpoints. All was quiet until I nearly stepped on a giant tarantula and instinctively shouted in alarm. Regaining my composure, I bent down for a closer inspection as it plodded across the surface. Its brown abdomen was the size of my thumb, and its thousands of little hairs almost perfectly resembled the thin needles of the teddy-bear cholla. It was so eerily beautiful and astonishingly alive, the first living animal I had seen since our arrival. *I bet this guy knows where the deer are*, I thought to myself.

As we came closer to the mountain's nucleus, the wash narrowed into a canyon, leading us through a series of dry waterfalls. Sand-filled pools spread beneath boulders polished smooth by past flows, creating natural staircases where waterfalls would cascade once the rains began. Climbing through a dry waterfall has a surreal effect on the imagination, making it impossible not to envision how water would change the experience. During a light summer rain, these falls would provide a picturesque setting in the form of shallow pools perfect for cooling off in triple digits. During a sustained monsoon drenching, the sudden force of water would knock down anyone trying this climb and spit them out of the canyon with relative ease.

A mile onward, and three hours after we started up the wash, the terrain started to flatten out. Behind us, the wash slowly fell six miles back to where we had started. Ahead, the wash descended for another six miles, coming to its abrupt end at the Colorado River. But in the mountain's glorious epicenter, the flatness extended out for a quarter mile in every direction. Mountain peaks surrounded us on all sides, and entering the flat felt like being in the center of a bowl. I turned around to see where we had hiked in from, but I saw only a wall of sharp rises, without an entrance in sight, as if we had emerged out of nowhere.

The scene was unlike anything I had ever witnessed in this parched country. It felt like a xeric version of Eden, a desert meadow of life and color. The spring was tucked into a slot canyon on the other side of the meadow, and thanks to dense trees and steep crags, pockets of shade abounded in which water could linger and nourish a profusion of flora. I sighted barrel cactus, buckhorn cholla, catclaw acacia, desert ironwood, honey mesquite, beavertail cactus, and tall, spiny ocotillos, some of which carried a flicker of flaming red flowers at their tips. There were boulder outcrops along the sandy floor and tiny caves hidden in the hills. And, for the first time all day, I heard the sound of birds. They swooped and sang and generally seemed baffled by the appearance of two humans walking in their territory.

The combination of the spring, the abundant shade, and the other elements that made this area a potentially suitable habitat for desert mule deer gave me optimism that we might finally see some of them. I became so optimistic, in fact, that I wondered if we'd see other animals too, perhaps a bighorn sheep clicking along a steep slab of granite, or a desert tortoise under the shade of an ironwood. My focus sharpened. I stayed close behind Aaron, and we moved in stealthy unison. If the hope of finding deer was measured in heartbeats per minute, I had enough for both of us. I absolutely believed we were going to find the crafty deer at any moment.

For another ninety minutes, we quietly roamed and glassed around every visible square in the grid we had visualized for the meadow. Nothing.

Aaron found a small boulder, sat down, and pulled food from his pack. He looked at me, defeated. "This is wild country, brother," he said slowly, at full volume, as if he had finally surrendered to the odds.

I sat down next to Aaron and felt my mind and body crashing back to earth. We were spent and still had another six miles back. Sand already has no regard for the walker, but that physical strain combined with the adrenaline of expectation had depleted my energy in new ways. We swapped snacks back and forth and traded sweat-drenched T-shirts for fresh ones. "Now, that was an experience," I offered in a simultaneous state of deep reverence and crestfallen disappointment.

With reality setting in, Aaron mused over the probability of his meat crisis continuing. "It's definitely not the first time I've come up empty," he said through laughter, throwing his hands in the air. But then he opened up, his tone reflective. "Even though I can't bring meat back to my family, it's all still worth it. Being out here in this desert and having all of this solitude and quiet . . . I end up spending so much time just thinking about the kids and my marriage and what they need and how I can be better. I always leave with more peace of mind and more focus than when I arrived."

Snaggletooth Rock

We eventually made our way out of the meadow, dropped down through the dry waterfalls, and followed the wash as it led us out of the mountains. Our pace quickened as the sun fell over the horizon. Back at camp, elk burgers sat in the cooler, awaiting a fire.

---

After two sunsets, two sunrises, and seventeen miles of traversing with scanning eyes and quiet steps, we didn't see a single mule deer. Although I never witnessed the culmination of a hunt, the intense sensory experiences and the rush of being an active participant in nature's intricate dance were both enlightening and transformative.

Rather than simply observing the plants, canyons, springs, and groves I encountered and then moving on, I began seeing them all in relation to the deer. The mesquite plant offered food, the canyon provided cover, the spring offered moisture, and the trees provided shade. Hunting required a complete ecosystem awareness. While my previous pilgrimages to the desert had helped me slow down and notice my surroundings, hunting had helped me understand the delicate interconnectedness of the biotic community and the diverse elements shaping these habitats. Hunting revealed itself to be more than a pursuit; it was another point of attachment between humans and nature.

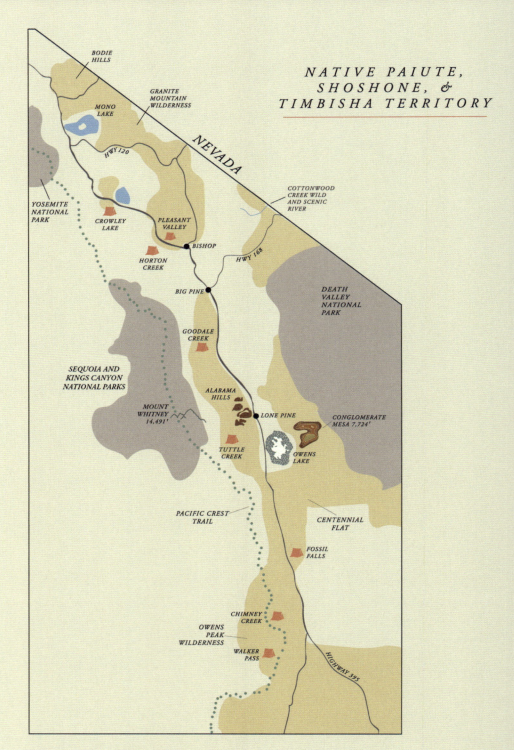

# EASTERN SIERRA: *There's Gold in the Hills*

*We were stuck.* Absolutely, white-knuckle, bare-bones stuck.

To say we were in the middle of nowhere would have been an understatement. The closest town to our west was a full day's walk. To our east, the nearest outpost was in Nevada, a marathon and a half away. And thanks to smoke from the 2021 Tamarack Fire blowing in from the north and the August daytime heat in the mid-80s, a long walk in either direction would have been a dreaded affair.

To make matters worse, dusk was coming to a close and a new-moon phase left us trapped in a state of complete darkness. Our phones, without reception, were reduced to flashlights.

It all happened on a gravel back road that proved to be a meandering path that tapered as we progressed down its length, like a funnel slowly squeezing us in. On one side of the road was a sandy streambed, and on the other were short hills of sagebrush and scattered rock formations that rose sharply in sporadic bursts.

We were headed toward a dispersed camping area I had found on an old map. Called Halfway Camp, it seemed like a promising setting for taking in the dark night sky. But as the road narrowed and transitioned from gravel to a sandy single lane, my anxiety spiked and I feared the terrain might be too challenging for both my two-wheel-drive van and my nerves.

Slowing the van to a crawl, I glanced over at my traveling companions, Sam and Paul. We had shared countless outdoor adventures, and among the three of us they were usually more optimistic about sketchy roads. For a few minutes, we debated whether to turn around or keep going. In the end, I won out, purely by virtue of being the driver.

This was not the first time I had ended up on an unpredictable road exploring BLM lands. After a few close calls on earlier trips, I decided to draft a set of ground rules to ensure future visits were smoother and safer. These rules included packing a shovel and several 2x6 boards (to help with stuck wheels) and following through on a sworn promise to never venture along an unknown road into an unknown campsite under the veil of night.

Adhering to these rules had steered me well through subsequent trips, and I had no intention of deviating now. I puttered on for a few hundred feet until I found a patch of dirt adjacent to the road that appeared large enough for a turnaround. I shifted into reverse and steered sharply, as if pulling into a tight parking spot. And that's when it happened—a sudden jolt upward. The emergency tire, which was stored under the back of the van, had wedged onto a boulder, lifting the back of the vehicle completely off the ground by several inches. I heard the wheels spinning in vain, as helpless and useless as I suddenly felt. We jumped out, crouched under the van, and faced the predicament: The van was now a tripod, balanced on its two front wheels and the boulder in the back.

The only thing to do was dig.

Beneath one rear tire, we placed two 2x6 boards to bridge the gap to the ground, and on the other side we employed a jack to hoist the van enough to free it from the boulder. It was still tilted at an awkward angle, but at least we had a plan. Using my large shovel, we took turns digging and jabbing and slicing around the boulder. For two long hours, we dug into the night, our rhythm slowed by the clumsiness of shoveling in a confined space. And then, ever so slightly, the boulder moved. Our victory celebration dissolved rather quickly, however, as we now faced the task of getting the beach ball–sized boulder out from under the van.

With whatever strength we could still muster, we haphazardly dug a long trench and then worked at pulling and prodding the boulder down the trench and away from the van. An hour later, we were free.

Hoping our embarrassing hiccup would go unnoticed by future travelers, we backfilled the ditch, returned the boulder, and got on our way. Having learned our lesson about turning around on such a narrow road, but also still committed to following my rule of not entering an unknown campground after dark, we accepted our fate: The only way out was backward. With Sam and Paul walking along the road behind me and yelling out instructions, I carefully drove in reverse until we reached a crossing with another dirt road. By 10:30 p.m., after finding a more suitable place to camp, we were at last sitting in our chairs under a dark sky sanctuary, breathing easy.

*Welcome to the Bodie Hills*, I thought to myself, shaking my head in disbelief.

EASTERN SIERRA: THERE'S GOLD IN THE HILLS

Bodie Hills

Rising directly north of Mono Lake in the Eastern Sierra, the high-altitude landscape of the Bodie Hills serves as an important ecological transition zone, bridging the Sierra Nevada range in the west with the Great Basin Desert in the east. Taking precedence in the center of this area are Potato Peak and Bodie Mountain, which both soar above 10,000 feet. To the east are Beauty and Bald Peaks, hovering around 9,000 feet and straddling the California and Nevada border with delicate charm. Despite their considerable heights, all of these mountains appear as if their tops have been lopped off, leaving the summits smoothed over, an appearance that earned them the name "Hills."

Sagebrush dominates the region, covering the slopes, meadows, and ridges. I can't help but wonder if the ubiquitous plant contributed to this 139,740-acre patch of land being overlooked by other agencies before ultimately falling under the BLM's jurisdiction. When compared to the iconic backdrops and forested abundance of Yosemite National Park, just a dozen miles to the west, the sea of sage here must have seemed inferior.

The other half of the Bodie Hills name is inseparably tied to an easterner named W. S. Bodey, who discovered gold in the region in 1859, sparking a series of short booms and long busts in the second half of the nineteenth century. At peak gold production, from 1878 to 1882, upwards of five thousand people lived in the town of Bodie (legend has it a sign painter misspelled Bodey and the name stuck), and today remnants of the ghost town, which include 114 buildings and an ore-crushing stamp mill, can be witnessed at Bodie Historic State Park, which encompasses 495 acres in the center of the Hills.

In the early years of the California gold rush, prospectors like W. S. Bodey went out in search of flakes and nuggets in streams, washes, and gulches. Mining was happening only on a small scale, meaning any individual with a pickaxe, shovel, and pan could mine for gold. As the influx of miners increased, these easily accessible "placer" deposits were quickly depleted, meaning miners were required to adopt more sophisticated methods of extraction. Soon, dynamite was being used to blast dangerous tunnels and shafts that led to gold veins running deep underground. Ore-crushing stamp mills were then employed to extract the gold from the surrounding rock.

EASTERN SIERRA: THERE'S GOLD IN THE HILLS

Mount Biedeman Wilderness Study Area

Bodie Hills

In the late nineteenth century, Bodie miners adopted a common technique called cyanide heap-leaching. Instead of blasting more tunnels, they pulverized the already-removed bedrock and placed it on leach pads, which are large impermeable liners that prevent cyanide and other chemicals from seeping into the soil during the gold-extraction process. A cyanide solution applied to the ore would separate any microscopic gold particles from the rock, and then the leftover materials were dumped in a tailing pond, which unfortunately posed a long-lasting environmental threat, as toxins, including mercury, often leached into the surrounding soil and watersheds. The cyanide heap-leaching method is still favored in the developed world today, but the scale, methods, and impacts are of course much larger than what was being done in Bodie. Even after more than a century of inactivity in the ghost town, the scars of mining remain visible on the surrounding land. Satellite photos reveal huge mounds of waste rock and so many tunnel shafts that the landscape resembles Swiss cheese.

---

The mineral that sparked the great rush to California in 1848 and catapulted the region into statehood still runs through the veins of the Bodie Hills. Recently identified gold deposits sit along the western boundary of the Walker Lane, a notable geological fault line along the California–Nevada border that is opening countless pathways for new mineral formations.

Driving this resurgent hunt for gold are two main factors: the dramatic rise in gold prices, which have increased by more than 600 percent since the dawn of the twenty-first century, and the facilitative framework of the General Mining Act of 1872.

The Mining Act, a cornerstone of US mining legislation, opened for exploration and acquisition all valuable mineral deposits on public lands. The antiquated law still allows individuals and corporations to file claims on twenty-acre parcels, without limits on how many parcels can be taken. While the financial cost to stake a claim is notably low, with claim fees totaling just a few hundred dollars, the law's most critically outdated aspects are the lack of leasing costs and the absence of royalty payments to the United States government for extracted minerals.

Another glaring omission from the initial act was its failure to include environmental protections and/or requirements for land rehabilitation, a choice that has led to widespread ecological damage. According to the California Department of Conservation, of the estimated forty-seven thousand abandoned mines in the state, 11 percent are still likely causing water and soil contamination. Although subsequent federal regulations have attempted to address these issues, they have largely been inadequate, often leaving taxpayers with the financial burden of remediation.

Thanks to the glaring shortcomings of the one-hundred-fifty-year-old Mining Act, both domestic and international corporations are eager to capitalize on the low-cost land grab. Through the efforts of the Bodie Hills Conservation Partnership, a broad coalition of groups actively fighting mining claims in the area, I learned about the many threats facing the Bodie Hills today. One of the corporations targeting the area is Paramount Gold, which, just in the month we were visiting, acquired claims on 2,260 acres just over the Nevada border. In a report from the Junior Mining Network, the president of Paramount, Glen Van Treek, spoke candidly about the prospects for their Bald Peak gold project: "All indications point to the potential for an open-pit deposit."

Conglomerate Mesa

Alabama Hills National Scenic Area

Highway 120

Conway Summit

While advocates argue that the gold mining industry contributes taxes, creates employment, and provides a crucial resource for modern technological devices, the reality isn't so rosy: Their taxes are minimal, their jobs often fleeting, and only 7 percent of the world's gold production is used for technological purposes. Although gold is used in consumer electronics, medical and aerospace devices, and other equipment and technologies, this usage is disproportionately small compared to the volume of gold extracted from the land. The majority of gold is allocated to jewelry (46 percent) and investments (47 percent), with governments and central banks stockpiling gold bullion, predominantly in underground vaults. This scenario underscores a profound irony: Almost half of what we extract from the ground goes right back into the ground.

Alabama Hills National Scenic Area

For many gold mining corporations, their practices can be distilled into a simple formula, as evidenced by a century and a half of environmental destruction and abandoned mines across the West: Find the mineral, devastate the landscape, take the money, and then abandon the site once profits inevitably dwindle. This search-and-destroy approach prioritizes profits over the environment every time. The aftermath is a scarred landscape like the one surrounding Bodie, an area marked by fragmented habitats, corroded machinery, crumbling structures, no-trespass fencing, and pervasive toxicity from mercury and acid.

When viewed from the perspective of today, the harsh reality of boom-and-bust gold mining becomes clear: The brief boom benefited only a fortunate few. The bust was for everyone else.

Bodie Wilderness Study Area

Beauty Peak

PRONGHORN ANTELOPE

*Antilocapra Americana*

The day after getting my van stuck, Sam, Paul, and I headed to the southeastern side of the area, closer to the site of Paramount's new claims. After a previous solo trip was cut short by the unanticipated arrival of a May snowstorm, I had been eager to return for a fuller experience. I also wanted to better understand what an open-pit deposit might mean for this part of the Hills. What flora and fauna thrived there? How would the ecosystem be affected?

With some initial help from Friends of the Inyo, a local conservation organization familiar with the terrain, I mapped out a long hike that would have us traverse the ephemeral Dry Lakes Plateau, pass near Beauty and Bald Peaks, and eventually lead us to the intersection of Atastra and Rough Creeks. The territory we were entering was largely unknown to hikers, and outside of a few beaten-up back roads, there were no trails. Luckily, walking on the high plateau is more like what I call "off-roaming"—a quieter, slower, cleaner, and more arduous form of off-roading. The beauty of walking in sagebrush country is there are no dense forests or unscalable mountains to hold you back.

We started our walk mid-morning, after several rounds of coffee. Our path started by followed a steep incline along a battered dirt road that would have wreaked havoc on almost any vehicle. Potholes were the size of wheelbarrows, and smooth undulations of mud resembled short snowdrifts, as if a winter of snow, wind, and ice had been followed by a summer of dry heat that had sculpted the road into a series of obstacles. Just as the terrain started to level out, we got our first glimpse of the Dry Lakes Plateau. The smaller of the lakes was a peeling, cracking, salty spectacle with tiny crimson plants spreading across the lakebed, resembling tentacles in a pale sea. We walked right to the middle of it, imagining what it would feel like in late spring to take a cold plunge in the exact same spot. The second lake was so vast it seemed to run straight off the horizon. Lingering smoke from the Tamarack Fire painted the sky a hazy pink.

Looming over the lakes was Beauty Peak, and I could see that it was aptly named. There was simply no other way to describe it. The gentle slope from the lake to the top of Beauty was ever so gradual, like the perfect sledding hill, and its double summit appeared rounded with the softest touch. The saddle between the summits was as inviting as any I had ever seen—an ideal spot for a tent or a lawn chair if you felt like hauling one up there.

As we admired Beauty Peak, Sam quietly gestured toward something in the distance that stopped us in our tracks. Sandwiched between us and the slow rise of Beauty was a herd of twenty-four pronghorn antelope. I had read they could sometimes be found here in late summer, but such occurrences were rare. Yet there they were, standing frozen among the sagebrush, staring straight at us. At least three males were in the group, their charcoal-colored horns strikingly juxtaposed against their white-and-tan fur coats. It was not lost on me that we were witnessing the second-fastest land animal on earth and the fastest over long distances. They can sustain speeds of thirty-five miles per hour for four straight miles. As we quietly crept toward them, they suddenly took off in a burst of collective energy, their long legs moving away from us in perfect unison. They darted this way and that, like a school of fish, until they disappeared over a short ridge and out of sight.

Though we were only a few miles into our walk, already the surprise of my surroundings was overwhelming. The surreal lakebeds. The pink skies. Beauty Peak. The pronghorn. We were left in one of those mindless trances that comes when nature completely engulfs you. No amount of planning or research can prepare you for an experience like that, and this one seemed to hit the three of us simultaneously as we forged our own paths across the long expanse of the second dry lake, contemplating the experience each in our own ways.

We reconvened along the old road, now transformed into a two-track path with vegetation sprouting from the center strip. A small herd of cows welcomed our arrival with indifference, barely looking up as they meandered through the maze of sagebrush. Our pace was relaxed, the late summer sun offering plenty of daylight for the kind of unhurried walks I had grown accustomed to. Under the outstretched canopy of a lonely juniper, we stopped to consult with my offline maps for the best route to the intersection of Atastra and Rough Creeks. From where we stood, the road took a decided turn downward, dropping six hundred feet into a small meadow, where it abruptly came to an end. The end of the road meant we would be on our own to the creek intersection.

While Paul and I deliberated, Sam had found a noteworthy chunk of obsidian nearby, and we passed the black igneous rock around in admiration. This volcanic glass, once widely utilized by Native people for tools and trade, served as a tangible connection to those who roamed these hills long before us, and before the cattle, the ranchers, and the forty-niners descended upon the landscape.

This is the traditional homeland of the Kootzaduka'a, the southernmost band of the Numu People, who have long called the Mono Basin home and once moved freely between the Bodie Hills, Mono Lake, and the Sierra Nevada.

The Kootzaduka'a people's yearly cycle through their various homelands was closely tied to the availability of regional food sources at certain times of year. The severe winters and the unpredictable timing of snowmelt made summer the most active period for their relocations, when the men would venture into the high Sierra to pursue deer and bighorn sheep while the women engaged in the meticulous work of seed harvesting in the meadows and along creekbanks. They focused on the seeds of nutrient-rich bunchgrasses, such as giant wildrye, Indian ricegrass, and desert needlegrass, essential components of their diet.

As the summer waned, the Kootzaduka'a would migrate to the northern shores of Mono Lake, where they held rabbit drives and harvested kutsavi, the larvae of brine flies. And growing on the southern slopes of the Bodie Hills were pinyon pine trees, which contained the pine nuts that would help sustain the community through harsh winters. In *My First Summer in the Sierra*, published in 1911 but based on his experiences in 1869, a young John Muir described a fall nut harvest he witnessed among the Kootzaduka'a: "Old and young, all are mounted on ponies and start in great glee to the nut-lands.... Arriving at some well-known central point where grass and water are found... the men with poles ascend the ridges to

Dry Lakes Plateau

Granite Mountain

Chimney Peak Wilderness

the laden trees, followed by the children. Then the beating begins right merrily, the burs fly in every direction, rolling down the slopes, lodging here and there against rocks and sage-brushes, chased and gathered by the women and children with fine natural gladness."

But the discovery of gold in Bodie changed everything for the original stewards. As the gold boom took over in the late 1870s, thousands of fortune seekers arrived, demanding resources to support their quest. Prospectors cut down trees to construct mining tunnels, and the influx of townsfolk took from the area what they needed for housing, food, and firewood. Simultaneously, the Homestead Act lured ranchers to the fertile areas around Mono Lake, and there they destroyed the Kootzaduka'a gathering meadows, converting them to sustain livestock and crops to meet the miners' needs. This agricultural takeover and mineral-extraction bonanza led to extensive logging of pinyon pines, not just for construction purposes but also to satisfy the insatiable need for cordwood, which fueled the stamp mills and kept the bustling town of Bodie operational.

As the four-year gold rush (1878–1882) faded with the dwindling gold reserves, the temporary community that had risen around it began to collapse. Bodie's volatile economy—all the shops, hotels, stage lines, mining and lumber companies, and individual prospectors and farmers—collapsed as the gold veins were exhausted. The town slowly fell into ruin, the mine shafts and tunnels were abandoned, and the surrounding hillsides, stripped of vegetation for mining and lumber, were left scarred and desolate.

The most disturbing tragedy, however, was the catastrophic impact on the Kootzaduka'a people, for whom the gold rush had spelled near extinction and the obliteration of their homeland and culture. Their important summer camps on the north shore of Mono Lake were now inhabited by homesteaders. The meadows and creeks where they collected bunchgrass seeds were now routes traveled by settlers, sheep, and cattle. And the pinyon pines, which nourished them through winter, were chopped down. With their homeland overrun, many Kootzaduka'a became laborers for the very people who displaced them and the very industries that destroyed their food sources.

The story of the Kootzaduka'a, much like those of numerous Native communities throughout the state and country, underscores the tragic legacy of gold mining. It reveals the troubling calculation in which fleeting profits are not just prioritized over landscapes but also deemed more precious than human lives.

ASPEN LEAF
*Populus Tremuloides*

After lunch beneath the juniper, Sam, Paul, and I followed the road down to the meadow. Beyond it, we encountered a dense tangle of trees and plants, with Atastra Creek quietly bubbling along behind. We picked our way downstream along a narrow cattle trail lined by encroaching sagebrush, wax currant, and rabbitbrush. To the east, crimson rock formations emerged from the earth, appearing out of place, as if they had been spewed from a distant volcano.

A half mile later, we arrived at the confluence where Atastra runs into Rough Creek. We couldn't help but smile. After a winter without much precipitation and a summer of aridity, this meeting point of the waterways was an unassuming affair; where they finally greeted one another, amidst willow saplings and swaying grasses, each creek ran only a foot wide. We huddled around the confluence and howled with delight at reaching our destination and getting to soak in the clear, cool water. We found grass along the banks of Rough Creek, discarded our backpacks, shoes, and socks, and sat down in the refreshing stream.

While Sam and Paul lingered in the water, I laced up my shoes and followed Rough Creek downstream until I came upon a small grove of aspens that had taken up residence in the confined spaces between the crimson rock formations lining the creek. Aspen groves are common around the Hills, sometimes in plain sight and other times tucked away in hidden valleys, and I was glad to see them here. I first fell in love with the tree when I learned of its revolutionary root system, which binds an entire forest together underground as a single organism. Every aspen comes from this root system, and each tree is a genetic replica of every other aspen within the grove—a complex system of clones in various stages of their life cycle. Their unusual bark, equipped with a thin green photosynthetic layer, enables the trees to grow in winter and provides important nutrients for migrating mule deer that run through the Bodie Hills, and the groves also provide crucial ecosystems and habitats for wildlife who use them for shade and visual cover.

As I stood among the aspens, I imagined the vast and interconnected system of roots just below the surface, all connected to one another, passing nutrients and information in one of the oldest dialogues we have on earth. A light breeze set the leaves quaking and trembling in a lively chorus, nature's very own tambourine, and the sound echoed off the canyon walls. I don't often feel as if nature is speaking to me, but in that moment I couldn't help but find a good place in the middle of the grove to sit down and listen.

In the late afternoon, we packed our things, dunked our hats in the creek for one last head bath, and retraced our steps. As we ascended the road and walked along the Dry Lakes Plateau, we noticed the wind had shifted and the smoky pink sky we had encountered that morning had given way to a brilliant blue. Beauty Peak emerged in stunning clarity. It was there, in Beauty's shadow, that the Bald Peak mining project would take center stage.

---

North of Beauty and just across the Nevada line, Paramount Gold has laid plans for exploratory drilling at eleven sites, with proposals to clear two additional areas for equipment staging and helicopter access. While these activities alone are poised to introduce a significant level of machinery and noise, a more pressing concern casts its own shadow with all the foreboding of a looming tragedy: What happens if their exploratory efforts uncover enough gold to justify a cyanide heap-leach mine? Looking out at the proposed site, it's almost beyond comprehension to imagine the quiet beauty of this landscape replaced with a giant open pit.

Helicopters would buzz overhead, delivering machinery and water for drilling. New access roads would be carved through the sagebrush. Razor wire fences would be constructed, blocking off hiking and hunting access and fragmenting wildlife corridors for the pronghorn and migrating mule deer. The dwindling population of bi-state sage grouse, which relies on these areas for mating season, would be further compromised. The remote quiet would be replaced by the constant roar of trucks rumbling between the open pit and the piles of waste rock. The smell of diesel and chemicals would infiltrate the clean air. And what about the potential for contamination in Rough Creek, which flows through canyons and aspens less than one mile away from the proposed site? The full repercussions of digging an open pit, accumulating piles of waste rock, and creating tailing ponds would irreversibly transform the environment.

The Bodie Hills were once deemed neither picturesque enough nor floristically diverse enough to hold value beyond the gold beneath them. But thanks to

Conway Summit

an expanding scientific understanding of ecosystems and biodiversity, the reality of the area is that it is an indispensable piece of a larger ecological puzzle—a refuge and corridor for dozens of fauna species, and a sanctuary for over seven hundred documented plants.

Time and again, it's these "leftover" BLM landscapes that face the greatest risk of destruction. Rough Creek, meandering through sagebrush, may struggle to rival the universally acclaimed scenery of Yosemite, and that difference comes at a cost. By preserving only the most spectacular places, we too often neglect these supposedly valueless BLM lands. Imagine the uproar if a mining operation proposed to drill on Half Dome? The public outcry would resonate worldwide, propelled by the millions of people who have formed a deep place attachment with the National Park. Yet it remains difficult to permanently protect unknown and unseen landscapes like the Bodie Hills.

---

Our dogged pursuit of gold has, worldwide and over many centuries, wiped out landscapes by steadily eroding the biodiversity of the natural world. The pressing question remains: When will we say enough? The Bodie Hills serve as a clarion call, urging us to draw a line, to choose preservation over plunder. In facing such a crossroads, we might do well to heed the timeless counsel Theodore Roosevelt delivered in 1903 as he gazed upon the Grand Canyon:

*Leave it as it is.*

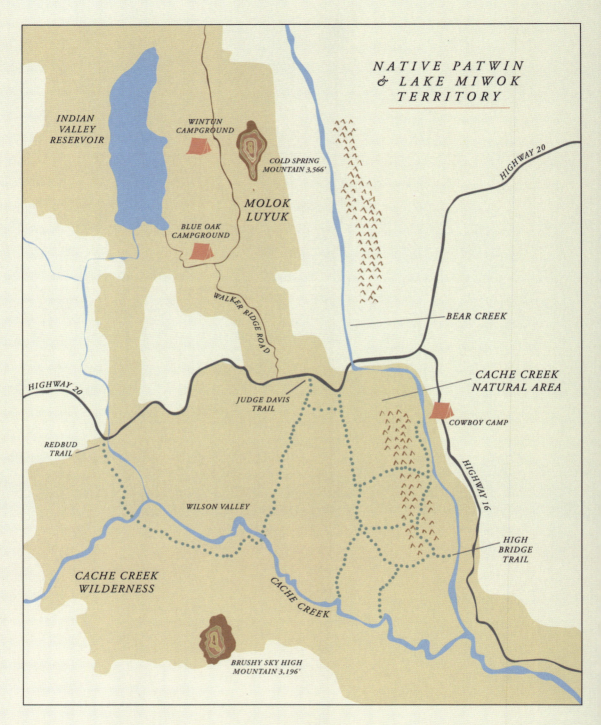

# BERRYESSA: *The Radical Center*

*We reached a small meadow* after driving a dozen miles on rugged gravel roads. It was the first week of November. My companions and I parked near a handful of colossal walnut trees that stood in wild disfigurement, their ancient branches dipping and twisting without a hint of uniformity. Each limb seemed to have sprouted in a chaotic search for the sun.

The meadow sits on the side of a mountain, forming something like a shelf, as if the mountain had briefly paused its upward thrust during the Jurassic period, taken a deep breath, flattened sideways, and then resumed its push toward the sky. The views from the shelf are vast, stretching down, across, and along the Inner Coast Ranges for a hundred miles. The flat-top 7,056-foot peak of Snow Mountain is easily sighted, as is the Indian Valley Reservoir some fifteen hundred feet below. The artificial lake is so still it appears almost frozen, like a sheet of azure-colored ice that fills in the lowlands. Foothill pines stand silhouetted against the sky, their thin, ghostly needles perfectly juxtaposed against their dark branches. Rising above the meadow is a steeply sloped hillside of dense manzanita that climbs five hundred feet until it reaches the 3,587-foot summit of Cold Spring Mountain.

A long, dry summer had bled indifferently into autumn, baking half of the flat to a golden crisp. The other half, surprisingly lush and green, flourished thanks to a natural spring that seeped from the namesake mountain. The cold water quietly dripped, trickled, and bubbled up from the earth, filling shallow hollows and slowly running downhill, nourishing a host of trees, plants, and grasses. Scrub oak, cypress, and toyon thrived along the edge of the saturated wetland.

The meadow, along with the surrounding hillsides and ridges, is called Molok Luyuk by the original inhabitants of these lands, which include people from the Yocha Dehe Wintun Nation, descendants of the Patwin. Yocha Dehe literally means "home by the spring water," and as you stand by the vibrant meadow, the connection they have to this place becomes clear.

Molok Luyuk

*TULE ELK*

*Cervus Canadensis Nannodes*

Molok Luyuk

In the Patwin language, Molok Luyuk translates to "Condor Ridge," a nod to the mightiest and rarest of birds that once soared over these hills in abundance and whose absence marked it now. In a letter to President Biden, tribal chairman Anthony Roberts spoke eloquently about the plants and animals within Molok Luyuk, describing them as "traditionally important to the lifeways of the Patwin people"—a reminder of the deep place attachment and interdependence the original inhabitants have with the ridge. Before colonialism devastated the tribes in different forms, the Patwin relied on local black bear, tule elk, and deer for sustenance, and the plants of the region were used for medicine and for materials to make objects for daily living: bows crafted from juniper, baskets from willow branches, and nets from milkweed.

The summit of Cold Spring Mountain is the highest peak in Molok Luyuk. It rises in the middle of the expansive 344,476-acre Berryessa Snow Mountain National Monument, whose name mirrors the geography surrounding it, with Berryessa Peak fifty miles to the south and Snow Mountain Wilderness the same distance to the north. The alignment forms a remarkable corridor stretching one hundred miles along the Inner Coast Ranges, with elevations ranging from near sea level to 7,000 feet. The US Forest Service oversees the northern half of the Monument, while the Bureau of Land Management administers the southern portion, including Molok Luyuk.

Cache Creek Wilderness

Wilson Valley

My companions for this trip (number twenty-eight!) were Andrew Fulks and Bob Schneider, longtime friends and passionate activists with backgrounds in environmental stewardship and geology. They both reside in the university town of Davis, not far from our destination. While researching for my second trip to the Berryessa region, I kept coming across references to them as central figures in the fascinating origin story of the National Monument. What intrigued me most was their unconventional approach during the process of securing the Monument's designation. I reached out to Andrew, hoping for a brief meeting over coffee to glean insights from his experience, but to my surprise, he proposed a day of hiking and conversation in the Monument itself. Our rendezvous point was Cowboy Camp, a serene spot along Bear Creek where I had chosen to set up camp for the week.

After the briefest of greetings (no time wasted for these guys), Andrew, Bob, and I left Cowboy Camp in Andrew's 1998 Dodge Ram pickup, a beast of a truck that affords him access to the roughest roads in the Monument. He calls the 4x4 his decadent pleasure and earnestly jokes that he makes up for its gas-guzzling engine by riding his bicycle to work every day.

As we bounced along the uneven road toward Molok Luyuk, Bob gave me a crash course in the science etched into the Monument's geology. At age seventy-five, he carried himself with an equal measure of vitality and curiosity, and his love for the geology in the Berryessa region practically spilled out of him. "This is one of the best places in the world to see plate tectonics," he told me with a grin. "It's either here or the Mariana Trench. And that's harder to get to."

After logging a three-mile hike and making numerous stops to inspect rocks, we arrived back at the meadow ready for lunch. Under the shade of the untamed walnut trees, sustenance came in the form of granola, salami, and crackers. Bob stood beside the truck as Andrew sat on the tailgate, legs dangling, and they traded stories about the landscape stretched out before us. Though ninety miles from home, they spoke of these public lands as their "backyard," and it was easy to see why. It seemed as if each ridge, canyon, and mountain was imbued with tales of their past adventures.

Andrew's gateway to the Berryessa region came through hiking on a few incomplete and overgrown trails that existed in the mid-1990s. Due to the general lack of access to the public lands, he decided to create a website to document where the trails were and how to access them through guides and maps. His work included restoring old trails and establishing new ones, and, true to form, he also brought upwards of four hundred people a year along with him on hikes. "We figured if people were able to get into the mountains," he told me, "they would be more likely to help protect them."

One of the longest trails he helped establish in the Monument was the fourteen-mile Berryessa Peak Trail, a difficult out-and-back route that required a climb of almost four thousand feet to reach the summit. The trail took three and a half years to complete and necessitated an exhaustive effort on the policy front, requiring approval from counties, the BLM, and a rancher who kindly granted a conservation easement so hikers could walk through their private property. But the hardest part, he said, was the physical work of actually "building the damn thing." Andrew and a small group of dedicated volunteers climbed the mountain season after season,

*MOLOK LUYUK*
*Berryessa Snow Mountain National Monument*

hauling loppers, hand saws, and shovels, until it was finally finished. In the end, the easement opened up nearly ten thousand acres of BLM land that had previously been cut off from access by the private properties surrounding Berryessa Peak.

After lunch, we inspected the lush half of the meadow, plant by plant, while our shoes sloshed in and out of puddles. I snagged a small branch from a cypress and crumbled the waxy, resinous leaves in my hand. They smelled overwhelmingly of gin and citrus, equal parts orange and lemon. I picked a few dead branches off the ground and stuffed them into my backpack, imagining their burning scent would pair well with an evening campfire.

As we walked from one half of the meadow to the other, from green to brown, from life to death, Andrew suddenly stopped in the middle of the dried-up grass and bent down in amazement. "Look at all this diversity!" he shouted.

I looked around, confused, thinking he had found something I couldn't see. My amateur naturalist abilities would have guessed we were in a monoculture of invasive cheatgrass.

"What are you seeing?" I asked.

He crouched low and showed me the seeds and stems of native purple needlegrass, vinegarweed, gumplant, and narrowleaf milkweed.

"I really should have brought some bags for seed collecting," he muttered while I furiously wrote in my notepad everything he pointed out. At fifty-one, Andrew had the energy of someone thirty years younger and the knowledge of someone thirty years older. He talked a mile a minute and knew as much about the littered brass bullet casings he cleaned up throughout the day as he did about the diversity of plants found in the meadow. As the sun began its tilt westward, we jumped back into the truck and drove deeper into Molok Luyuk. Riding shotgun, I asked Andrew and Bob to walk me through the backstory of how the Monument came to be.

---

The account they shared spans a dozen years and highlights the sheer audacity of the Monument's existence. Although both Andrew and Bob have a long history of individual conservation achievements, it was their collaborative effort that made the most significant impact. In 2002, they cofounded the nonprofit conservation organization Tuleyome to tackle the numerous challenges within the Berryessa region, including the mercury-polluted Cache Creek and the surrounding BLM lands, both of which were unprotected and inaccessible.

From the outset of the effort to secure Monument status for the area, Andrew, Bob, and their team of volunteers recognized that they would need to build a broad coalition of support. They diligently engaged with a wide range of stakeholders, including local landowners, nonprofits, Indigenous communities, recreationists, politicians, and regional businesses. They told me it was a long and bumpy road that included all the inevitable disagreements and frustrations, such as which public lands should be included in the Monument. Some surrounding counties supported the Monument, while others were strongly against it. At the higher level, it took a lot of work and skill to navigate the complex bureaucratic process of uniting various federal, state, and county agencies around the proposal.

They also made a concerted effort to be inclusive of people that had historically been at odds with Monument proposals. These voices included ranchers who owned land bordering the proposed Monument, many of whom came from families who were fourth- and fifth-generation farmers, folks with their own deep connection

Cache Creek Natural Area

to the land. Initially, they were leery of the proposal, fearing it would affect their grazing access, but they soon learned that the Monument designation would not alter existing grazing rights. Over time, many ranchers came to support the Monument, and today they continue to have a voice in management planning. Andrew reflected fondly on some of the friendships that were eventually formed with these families, visit after visit, year after year.

Remarkably, Tuleyome even garnered support from recreation users who often feel overlooked in conservation circles. Andrew emphasized his commitment as an "equal opportunity recreationist," asserting that public lands belong to everyone. This inclusive approach extended to OHV groups, a connection that then led to alliances with key figures such as Don Amador, then chair of California State Parks' Off-Highway Motor Vehicle Recreation Commission. Through numerous map-filled meetings and discussions over coffee, a compromise was achieved by allowing OHV use in some areas, while prohibiting access in areas of more critical environmental concern.

In an interview with the *LA Times* in January 2015, Amador reflected on the collaboration: "I believe we've moved past the era of feuds between environmental

Bear Creek

groups and off-road organizations. This stands as a model for conserving these spaces and finding common ground."

By the time the Berryessa Snow Mountain National Monument was designated by President Obama on July 10, 2015, after a full decade of building authentic relationships with those who would be most affected by its passage, Tuleyome had garnered the support of tribes, ranchers, sixty elected officials, hundreds of local businesses, recreationists, and nonprofits such as Latino Outdoors and Backcountry Hunters and Anglers.

Nine months after the official designation, a special ceremony was held at Cowboy Camp, and an estimated nine hundred people showed up for the festivities. Cecilia Aguiar-Curry, at the time the mayor of nearby Winters, opened the ceremony with these remarks: "The journey of this moment has been a story of dedication and persistence on the part of many local residents, business owners, outdoor recreation lovers, farmers, ranchers, community leaders, [and] sportsman and conservation groups."

The gathering included the notable officials and politicians involved in the designation, but also present were many of the unheralded individuals and groups who had helped make it a reality. Elementary-aged children held painted signs and banners, equestrian groups and ranchers stood alongside their horses, and OHV enthusiasts rode in on their lifted rigs. In a photo snapped at the ceremony, Bob and Andrew blend in with the crowd. Andrew is crouched on the ground, hands clasped together and wearing a joyous smile. Bob is standing near the back, half hidden behind others. The word he used to describe what he felt that day? "Grateful."

---

The transition of this area from unprotected BLM property to protected National Conservation Land not only curtailed future industrial threats, it also led to further research on and protection of the cultural sites, plants, animals, waterways, and biodiversity of the region. And in 2024, when the 13,696 acres in Molok Luyuk were added to the Monument, the plan called for co-stewardship between the BLM and the Yocha Dehe and Kletsel Dehe Wintun Nations, a collaboration that brings back to the land the traditional knowledge of those who have an intimate, multigenerational understanding of this place and its interwoven landscapes.

In terms of recreation, the Monument triggered a series of enhancements that continue to resonate in the communities surrounding Berryessa. There's now funding for comprehensive management plans that draw on expertise from various fields to ensure a balanced approach between conservation and recreation. This influx of resources also translates into improved access, educational initiatives, signage, maps, brochures, and funds to hire critical staff.

During my ten days in Berryessa, split between a spring trip and a fall trip, I reaped the benefits of the recent improvements. The BLM lands, once largely inaccessible and devoid of hiking trails, are now easily navigable. Thanks to the Monument's designation, anyone can now download an incredibly detailed BLM map and brochure that highlights the hundreds of miles of hiking trails, newly developed campgrounds, and a host of opportunities for hunters, anglers, boaters, equestrians, and OHV enthusiasts.

The story of the Monument's establishment offers timeless lessons about what it takes to protect BLM lands in the future: unwavering determination, a spirit of compromise, and a commitment to radical inclusivity.

Andrew's ethos as an "equal opportunity recreationist" resonated strongly with everything I had learned thus far in my travels. Perhaps the most salient message was that everyone's path to the outdoors is uniquely their own. Mine, like Andrew's, had begun with countless steps over miles of trails. For Bob, the connection was through mountaineering and geology. For Don Amador, it came on the back of a motorized utility vehicle. Back in the Mojave, for Aaron and Ryan, it unfolded through the lens of hunting, and for Emma in the Carrizo, it was through the study of flowers. Acknowledging and valuing these varied avenues toward forming a relationship with nature will be crucial as we look to the uncertain future of landscape and wildlife conservation. As we collectively grapple with a warming planet, the accelerating loss of species and biodiversity, and the severe imbalance between extraction and conservation on BLM lands, it becomes clearer than ever how necessary a broad coalition of nature advocates is to achieving our goals.

This principle of inclusivity in conservation is known as the "radical center" approach. In Randall K. Wilson's 2014 book *America's Public Lands*, he defines the radical center as "the idea that compromise represents an audacious but critical approach to problem solving in the age of zero-sum politics." As partisan ideologues increasingly push the American public further away from each other and

Cowboy Camp

out toward the extremes, the new conservation radicals emerge, paradoxically, from those positioned closer to the middle.

The widespread support Tuleyome assembled for the designation of the Berryessa Snow Mountain National Monument echoes the success of conservation programs in the 1960s and '70s, a period marked by the passage of significant environmental laws with overwhelming bipartisan backing. Examples include the Wilderness Act (passed 374–1), the Endangered Species Act (passed 390–12), and the BLM's own Federal Land Policy and Management Act (passed in the Senate 78–11). Even the Clean Water Act, which was initially mired in lengthy debates, eventually found resolution after ten months of intense negotiations. Described as "a product of torturous agreements" by the *Congressional Quarterly*, the bill ultimately passed the House by a vote of 366–11 and received unanimous approval in the Senate, 74–0. President Nixon, in a 1971 address to Congress, put his stamp on this era of radical center conservation by eloquently calling our public lands the "breathing space" of the nation and emphasizing the importance of safeguarding their environmental values for future generations.

---

After a full day navigating our way through Molok Luyuk, we returned circuitously to Cowboy Camp via Bear Valley, completing our fifty-mile loop. I shook hands with Andrew, thanking him for his time and generosity. Bob offered a hug, which I gladly accepted. As the soft light of late afternoon enveloped this little corner of their backyard, I waved goodbye as they pulled out of camp.

I stood there for a while, lulled by the stories embedded in the landscape, wondering if the true monument was not just the land but also the collective memory and unwavering commitment of those who cherished it. With a heart full of gratitude for all of those who made this experience possible, I built a fire and sank into my camp chair. I watched the last shred of daylight disappear in the west, but not before it perfectly illuminated a lone blue oak standing on a distant ridgeline.

By 6 p.m., darkness had completed its takeover. The stars were doing their best, but so were some thin clouds that hung like sheets of muslin, shrouding the night sky. A coyote to the north howled for a while. Then another one to the west started up. They alternated for a few minutes and then both of them called it quits. The only consistent tones came from Bear Creek, fifty feet away, but it ran quietly this late in the season, a gentle murmur that soothed the senses.

By 7 p.m., the crisp air carried the beginnings of winter. I shifted my legs from one side of the flames to the other and threw my harvested cypress branches on the fire. Hints of citrus slow-danced with the smoke and wafted through the air. At 7:53, I crawled into my sleeping bag.

I was out before the moon appeared.

# KING RANGE
National Conservation Area

# KING RANGE: *Reciprocity*

*There are three and a half roads* leading into the King Range National Conservation Area. I consider the southern route half a road because unless you have the kind of vehicle that can ford creeks, handle quicksand, and maneuver around boulders and fallen trees, you can pretty much forget about it.

The northern route is arguably the most beautiful, and at roughly two hours' drive time, it's also the longest. Beginning in the quaint town of Ferndale, the road takes a slow, meandering path through farms, meadows, and forests. Eventually, just as you feel as if the road is going to direct you straight off a cliff and into the ocean at the famed Cape Mendocino, the road makes a sharp left and sends you careening down toward the Pacific at the kind of sharp descent that would make a modern road engineer shudder. From here, you've got ocean views for several miles before turning inland, where you'll head past the tiny hamlet of Petrolia and then make a right on Lighthouse Road, back toward the Pacific, having reached the northern tip of the King Range.

The two main routes both come from the east, off Highway 101. The first begins in Garberville and sends you twisting along the southern portion of the King Range before you dead-end at the seaside village of Shelter Cove. The final option begins a few miles north of Weott and snakes through a thick maze of redwoods before traversing a steep mountain that eventually follows the Mattole River until you conclude your drive at the same place the northern route ends.

Whatever route you decide on to get to the King Range—and I can only say they are all wonderfully wild—you can expect some nausea. There will be inclines and declines, the kind that make your stomach rise and fall like on a rollercoaster, and there is an endless supply of hairpin turns. You can also expect potholes, pavement crevices the exact width of your vehicle's tire, and possibly the disappearance of entire sections of road. The various routes will take roughly 45 to 120 minutes, depending on how the aforementioned nausea and road conditions influence your speed.

The stark isolation of the King Range is the result of a number of natural factors and one fortuitous lucky break. The primary natural consideration is the weather. The area receives around one hundred inches of rain per year, which creates mudslides and swollen rivers that wreak havoc on road infrastructure. Frequent winds out of the northwest are relentless enough to regularly knock down trees and power lines. And then there are the numerous earthquakes, which jolt the area eighty times a year thanks to the region's close proximity to the infamous Mendocino Triple Junction. This is where the North American, Gorda, and Pacific tectonic plates all collide, making the King Range one of the most seismically active places on the planet. The severely folded and faulted mountains are in a constant state of rearrangement.

The fortunate break is that the iconic Highway 1—renowned for tracing the California coast for 655 miles from Orange County northward—abruptly veers inland when it reaches the King Range. The absence of a road along the Pacific at this point has serendipitously preserved the longest expanse of undeveloped coastline in California.

Romantics refer to this area as the Lost Coast, a name that conjures intoxicating and mysterious visions that call to my sense of adventure. The first of my many journeys there came in 2012, when with my wife and our two children, ages three years and three months, ventured in via the route from Ferndale. Despite only having one day to road-trip through its outer reaches, we couldn't believe the beauty we witnessed. We stopped frequently along the lonely road, over and over, to admire the moss-covered forests, splendid views, and vibrant tide pools. Little did I know at the time that hidden within the Lost Coast was a 68,000-acre parcel of public land called the King Range National Conservation Area, which was overseen, of course, by the Bureau of Land Management.

---

Geographically, the King Range National Conservation Area is a slender ribbon of mountainous land stretching along the Pacific Ocean from Sinkyone Wilderness State Park in the south to the mouth of the Mattole River in the north. The area is neatly framed by the free-flowing Mattole River, which meanders for sixty-two miles just outside the Conservation Area before coming to its inevitable end where the freshwater greets the salty currents of the Pacific Ocean.

Named for the King Range of mountains that rise sharply from the beach, this area is characterized by its dramatic topography. King Peak, the sentinel of the range, climbs precipitously from the Pacific to over 4,000 feet in elevation within just three miles.

In addition to the unique land features, the Conservation Area is also home to an array of wildlife, including black bears, mountain lions, black-tailed deer, and the endangered northern spotted owl, which takes refuge in forests of ancient Douglas fir. The Pacific coastline, peppered with islets, provides sanctuary to diverse avian species and elephant seals, and the Mattole River and its seventy-four tributaries are vital waters for coho and Chinook salmon. Thanks to the generosity of moisture, come winter and spring the area morphs into a lush temperate rain forest (reminiscent of those in the Pacific Northwest) that supports dozens of species of mosses, mushrooms (giant puffball and lion's mane!), other fungi (crystal brain and earpick!), and lichens (Methuselah's beard!).

---

The establishment of the King Range National Conservation Area was a process marked by tragedy and resilience. In response to the deterioration of the landscape due to rapacious logging practices, along with the lack of recreation opportunities, Clement Miller, a Democratic congressman from California's 1st District, introduced a bill in 1961 to establish the nation's first-ever National Conservation Area. The bill proposed a seismic shake-up to the BLM's near-exclusive focus on mining and grazing. Miller proposed that the vision should expand to include conservation and recreation. The groundbreaking ideology of the bill cannot be underestimated—it called for a multi-use plan of BLM lands *fifteen years before* the Federal Land Policy and Management Act was passed in 1976.

During Miller's earnest sixty-minute speech on the floor, he alluded to this shift in priorities: "The recreational field is a new one for BLM. The bureau has been a sort of stepchild among the government's great land management agencies for too long. I am looking forward to a Cinderella transformation with high expectations." Perhaps

Mattole Estuary

for the first time in the bureau's history, here was someone who envisioned BLM lands as more than just commodities to be exploited.

Despite the timely support of President Kennedy's plea for balanced land use, Miller's initial bill failed in committee. Undeterred, he reintroduced it in 1962, when it failed once again. And then tragically, on October 7, 1962, just one month before the upcoming election, Miller died in a plane crash on the way to a campaign event. One of the last places Miller would have witnessed from the sky was the very landscape he had spent his final years trying to protect.

In a remarkable twist, his political rival, Republican Donald Clausen, took up Miller's cause after being elected in his absence. Clausen brought the bill forward in 1963 and again in 1965, but both attempts were unsuccessful. In 1967, Clausen, along with two other proponents, pitched the bill five separate times in the House and Senate, only to face a clean sweep of rejections. But against all odds, Clausen's persistence finally paid off when, on the twelfth attempt, eight years after Miller's first proposal, the bill received a favorable report and passed through to a vote. On October 21, 1970, the shared dream was realized with the establishment of the King Range National Conservation Area.

---

Inspired by the legacy of this inaugural Conservation Area, I was eager for a return visit. After a twelve-hour long-haul road trip from Los Angeles with my friend Noah, we made the final approach via Garberville, arriving at the southern end of the Conservation Area just as the late summer sun was falling toward the Pacific horizon. Noah, an illustrator by trade, is the kind of cheery traveler who runs on an inexhaustible reserve of optimism and adaptability, an ideal partner for experiencing the remote landscape.

We parked near Black Sands Beach and made a beeline for the shore, hoping to catch the last of the magic hour before the stars took over. Low tide revealed a wide beach, and we walked northward for half a mile, our long shadows trailing behind us on the sand dunes. We paused only to watch the last sliver of sun drop out of sight. By the time we headed back, stars were filling the sky, multiplying by the thousands every time we looked up.

Black Sands Beach

    The Wailaki Campground, named after one of the local Indigenous tribes, is a small camp nestled a few miles inland and surrounded by a dense forest of towering Douglas firs. The South Fork of Bear Creek, which was quietly flowing in the late summer, runs directly through the camp. Thanks to the complete absence of artificial light amidst the remote landscape, driving into the campground felt like entering a cave. After a long day on the road, we blissfully settled into our camp chairs for a quiet night under the canopy of fir, the steady sound of Bear Creek perfectly paired with a glass of whiskey. We had made it.

    I had decided to spend the first half of the week at the southern end of the Conservation Area for easy access to the trailheads that opened up to eighty miles of routes crisscrossing the range. We didn't have permits to hike the world-famous Lost Coast Trail, which runs along the coastline for twenty-five miles between Black Sands Beach and the Mattole River, but we had plans for several day hikes throughout the week. We started with the Horse Mountain Trail, an eleven-mile trek that begins midway up the King Range, descends to the beach, and then climbs back to the trailhead.

Even on a warm day in September after the long dry season, stepping onto the trail felt like walking into a rain forest. Giant chain ferns drooped over the path, welcoming our arrival by running their soft edges against our arms and legs. The gnarled roots of Pacific madrones twisted out laterally along the ground like tentacles and then dove into the earth, anchoring themselves against the frequent tremors. Elderly bigleaf maples stood defiantly, and oh how they sang. Their namesake leaves, sometimes a foot in length, were like thousands of outstretched hands offering shade and good tidings. Mosses climbed and clung to their bark and low-hanging limbs, covering them in shades of fluorescent green.

Farther along our descent, yerba santa grew in abundant patches. Their long, thin leaves were so glossy and sticky they appeared wet, and I passed through enough of them to turn my pants tacky. Poison oak crept along the floor like a snake, climbing ever so slowly over roots and rocks and streams with no end in sight. I stepped and jumped my way around the allergenic leaves with precision. The charmingly named plant pearly everlasting lined the trail with clusters of white flowers and yellow buds. And the Douglas firs! Those venerable saints of the great forest, who have stood the longest and reached the highest, their crowns touching the Pacific breeze, their roots commingling with neighboring trees through an underground network of fungi that communicates in an ancient dialogue of chemicals, hormones, and signals—they were magnificent.

Wailaki Campground

Every so often, we briefly emerged from the foliage to take in the expansive panorama stretching down the mountain and out to the ocean. Distant firs and cedars stood silhouetted against the deep blue sea, and rocky islands popped up along the shoreline. Then, just as soon as we found the views, they disappeared once again.

We reached the beach as high tide crashed at the very edge of where the sand met the forest. Massive silver driftwood logs were stacked up haphazardly across the sand. Walking south along the beach, we arrived at the terminus of Horse Mountain Creek, where the water quietly exited a canyon and pooled near the shoreline. We peeled off our worn socks and shoes and waded through the crystal-clear waters.

As we navigated the descent, bears had been my constant concern, prompting me to stay vigilant to the sounds and movements within the dense understory. Another thought had me scanning the trees: The northern spotted owl, a federally threatened species, finds refuge in the King Range. Unfortunately, their numbers are rapidly declining, and a recent study found that between 1995 and 2018, spotted owl populations in many areas shrank by at least 65 percent. Initially, habitat disturbances like logging and road building were to blame, but these days the endangered owl was being pushed out by the barred owl, a more aggressive species that competes for food and nesting sites.

Horse Mountain Trail

Recognized as an "indicator species" across the Pacific Northwest and Northern California, the spotted owl only occurs in areas where the overall forest ecosystem is healthy and intact. When the owl was designated as threatened in 1990, the new status catalyzed protective measures not only for the northern spotted owl but also for other rare species, including red tree voles, northern flying squirrels, truffle mushrooms, and the old-growth Douglas fir stands they all depend on. This interdependence forms a vital circle of reliance and connectivity: Douglas firs create the cool, shaded conditions necessary for truffles, which are crucial for the diet of the northern flying squirrel, which in turn is the primary food source for the spotted owl.

King Range Wilderness

Halfway through our climb back up the mountain, the landscape off the trail dropped into a ravine, and as I tightroped along the tenuous ledge, I heard the loud crunch of breaking branches. I immediately stopped in my tracks, following the noise through a screen of trees until I laid eyes on the culprit. Momentarily paused in the brush on the opposite side of the ravine was a large black bear. I froze in silence, scanned for cubs, and attempted to judge the bear's intentions from its movements, which seemed agitated. For thirty long seconds, the bear and I stood motionless. And then it bolted away from me in great bursts, bulldozing through the woods until it was out of sight.

And then, unbelievably, a few miles later while waiting for Noah, I looked to the trees and found a northern spotted owl perched on the thin branches of a young California laurel. My jaw dropped. Seeing a bear was relatively common in the King Range, but encountering this elusive nocturnal owl in daylight was a rare treat. I paused, holding my breath, hoping it wouldn't move. Its fierce, dark eyes gazed straight at me. It was so beautiful, so delicate. The light creamy spots along its feathers and the matching horizontal brushstrokes on its chest made it look like a painting. When Noah arrived a few minutes later, we quietly stared together with happy smiles, marveling at the bird's perfected stillness and the luck of the moment.

After a few more days exploring the southern end of the Conservation Area, we took a two-hour drive along back roads to the northern end where the Mattole River greets the Pacific. The temperature difference between the coast and just a few miles inland can be downright drastic, and our experience was no exception. As we approached the Mattole Beach Campground, we were welcomed by thick fog, strong winds, and temperatures thirty degrees cooler than they were just ten miles inland.

The northern boundary of the Conservation Area is where the Mattole River begins widening as it draws closer to the ocean. During summer, when the river's volume isn't large enough to make it all the way to the Pacific, it backs up behind a long sandbar to form a freshwater lagoon. Sea lions and harbor seals doze and hunt along the sandbar that separates river from ocean, and river otters peek their heads in and out of the lagoon, seemingly unhurried. Flanked by Moore Hill to the north and Strawberry Rock to the south, both of which rise seven hundred feet on either side of the river, the setting is like a magnet for the senses. The mountains call your feet upward. The crashing waves offer a lullaby of sound. The warm lagoon waters invite you in for a swim. Upon arriving at the campground, we immediately headed straight for the lagoon.

KING RANGE: RECIPROCITY   *173*

Mattole Beach Campground

NORTHERN SPOTTED OWL
*Strix Occidentalis Caurina*

It was here along the final few miles of river that the last known speakers of the Mattole language lived. One of them, a man named Joe Duncan (his Native name was believed to be Taralesh), lived with his son Ike in a small cabin near the mouth of the river, and in 1927, a linguist named Dr. Fang-Kuei Li, famous for his work with Native American languages, spoke with the Duncans. Joe described plainly the fate of his people at the hands of American militias and settlers, telling stories that exemplified the campaign of genocide and enslavement inflicted on so many Indigenous people in northwestern California and elsewhere.

Duncan, born in 1850, recounted witnessing the violent arrival of white settlers during his childhood and the widespread killing of his people, including women and babies. To escape the violence, he fled to the Eel River, and while he waited for things to quiet down so he could return, the government took the Indigenous people's land and sold it to American citizens.

Looking out over the lagoon and imagining the Duncans' cabin standing somewhere within our range of view brought home the heartbreaking story of loss. Without knowing exactly how to reconcile the tragedy with my own experience as a visitor, I spoke the Duncans' names out loud. I mentally traced a line from my present experience in the lagoon to a moment back in time, imagining the long history of humans living on and shaping this landscape. Coming to terms with the way the present is inextricably linked to the past has deepened my attachment to these lands. The writer Robert Macfarlane has perfectly articulated the question that hangs over my mood: "The pasts of these places complicate and darken their present wildness; caution against romanticism and blitheness. To be in such landscapes is to be caught in a double-bind: how is it possible to love them in the present, but also acknowledge their troubled histories?"

For at least seven hundred years, the river and ocean have remained a revered symbol within the Mattole culture and way of life. This meaningful interdependence is illuminated in other rare interviews with the Native peoples of this area, as documented in the 1960 publication *Fishing among the Indians of Northwest California*, by Alfred Kroeber and Samuel Barrett. In it, anthropologist Gordon Hewes recounts insights shared by the Mattole about their reverent relationship with the aquatic environment. He writes, "Among the Mattole, conduct toward waves is prescribed: The water watches you and has a definite attitude, favorable or otherwise, toward you. Do not speak in passing rough water in a stream. Do not look at water very long for any one time, unless you have been to this same spot ten times or more."

Central to this relationship with the water is a relationship with the salmon, which plays a sustaining role in Mattole communities (now part of the Bear River Band of the Rohnerville Rancheria, a federally recognized tribe formed in 1910 from displaced Mattole, Bear River, and Wiyot people). The people's dynamic, mutual exchange with the salmon is rooted in a deep understanding of reciprocity. The ecological writer and philosopher David Abram explores this concept in his 2001 essay "Reciprocity." He writes, "Reciprocity, the ceaseless give and take, the flow that moves in two directions—this is the real teaching of the salmon. It is the foundation of any real ethic: . . . if you wish to receive sustenance from the land then you must offer sustenance *to* the land in return." Just as the Mattole people have long honored this reciprocal relationship with the salmon, the salmon themselves engage in a similar exchange with the land and sea.

---

Each year in late October, as the first autumn rains begin falling, a great gathering takes place on both sides of the sandbar. Coho and Chinook patiently bide their time, waiting for the river to develop enough volume and pressure to blow through the sandbar that has been sculpted by the wind and waves throughout the summer. When this happens, the lagoon will become an estuary.

On the Pacific side, adult salmon are waiting to complete their migration. Having spent the last several years traversing the continental shelf that runs up and down the Pacific from Point Reyes to Oregon, they are now ready to return to their spawning grounds. Once across the berm of the former sandbar, they will spend the last of their energy swimming upstream in search of their birthplace and then preparing the right conditions for the next generation of salmon to be born. Their return marks one of nature's greatest riddles. In "Reciprocity," Abram eloquently asks the questions that scientists have been studying for over a century: "By what

## COHO SALMON
*Oncorhynchus kisutch*

Coastal sand verbena

magic do they find their way back not just into the right river mouth, or the right tributary of that river, but into the precise little stream where they once hatched?"

Scientists believe the fish intuitively know how to get home through magnetic, olfactory, and celestial clues. Many tribes of the Pacific Northwest for whom their relationship with the salmon is part of their identity tell stories of salmon as immortals that take human form in the ocean. Each year, the salmon king orders those humans to clothe themselves in fish skins once again and return to the river as a gift to the tribes. Whether by intuition, holy mystery, or some supernatural enigma beyond our understanding, the salmon have returned to the Mattole.

On the inland side of the sandbar, meanwhile, adolescent salmon are navigating the estuary waters and waiting for the second phase of their life to begin. These anadromous species are genetically designed to move from the warmer fresh water of their birth to the cooler salt water where they spend the majority of their lives, fattening up on zooplankton, invertebrates, and eventually other fish species such as herring and smelt.

Prosper Ridge

Windy Point

In their passage from river to sea and back to river, the salmon act as a crucial intermediary between the terrestrial and aquatic realms. In other words, reciprocity between forests and oceans is miraculously carried out by the salmon themselves. Forests, enriched with nutrients, transfer these benefits to rivers, which the salmon then carry to the sea. Likewise, the ocean nourishes the salmon with vital nutrients like nitrogen, carbon, and phosphorous, and upon the fish's return to the river and eventual death in those waters, these nutrients are released, benefiting a wide array of forest dwellers—from bears and otters to eagles—that spread these contributions throughout the forest, enriching the soil and fostering plant growth. Abram captures this cyclical process by suggesting it's like "a kind of breathing": The forest exhales the salmon out to the ocean and then inhales the nourishment they bring back into the forest.

But just as the Mattole people's reciprocity with salmon was once severely threatened by colonialism, the salmon's own reciprocity with the land is now in grave jeopardy. The legendary salmon runs of the Mattole River, once a testament to the health and vibrancy of the ecosystem, have dwindled dramatically in the last half century. Land mismanagement and human-caused climate change have led to a severe reduction in the native populations of coho and Chinook.

The catastrophic decline in Mattole River salmon species can be directly tied to the rampant logging that once took place in the greater 304-square-mile Mattole watershed. Land ownership within the watershed consists of private landowners (46 percent), BLM lands (12 percent), and commercial timber companies, who own much of the rest. Since the first management plan for the King Range was implemented in 1974, logging in the area has been limited to two salvage sales that took place after wildfires, in 1975 and 1988. While that arrangement protected that 12 percent of land belonging to the BLM, it didn't stop logging elsewhere, and even then those protections came after decades of destruction. Between 1947 and 1987, an astonishing 82 percent of the timber in the watershed was harvested, and it wasn't until the 1994 Northwest Forest Plan was enacted—regulations that protected old-growth forests *and* the northern spotted owl that depend on them—that commercial logging was more closely regulated in the watershed.

But much of the damage had already been done. The staggering scale of logging meant thousands of miles of skid roads, which blocked streams and natural pathways for water to drain. When a pair of storms dumped thirty inches of rain in December of 1955 and then again in 1964, the denuded mountains turned into massive landslides that fell into the Mattole River and its many tributaries. The sudden rush of sediment filled in stream channels and wiped out bankside riparian vegetation, which had provided the shade and cooler water temperatures that are integral to the salmon's survival. The addition of frequent droughts, which lower flow rates and increase water temperature, has nearly put the nail in the coffin for the Mattole strains of coho and Chinook. If not for a grassroots campaign to save the salmon, they might already have vanished.

---

A beacon of hope emerged in the 1980s when individuals and Indigenous communities, galvanized by their shared resolve to revive the Mattole salmon and heal the watershed, formed organizations and intertribal alliances dedicated to the cause. Their four-decade journey is yet another unfolding story of reciprocity.

## PUNTA GORDA LIGHTHOUSE
*King Range*

The Mattole Salmon Group, founded in 1980 by local residents of the lower Mattole Valley, was led by Freeman House, a writer and commercial salmon fisherman. The group aimed to understand the fish's needs and create a strategic preservation plan, but they soon recognized that the salmon's struggles were part of broader ecological deficiencies that would need to be addressed. This revelation led to the formation, three years later, of the Mattole Restoration Council, whose mission is to restore and conserve the watershed's ecosystems. Four years after that, the Sanctuary Forest was established to collaborate with upstream landowners in protecting the old-growth forests near the Mattole headwaters.

These organizations began with detailed studies on fish populations and estuary health, including annual spawner surveys, egg collection, and documenting the decline in old-growth forests. The Mattole Restoration Council shared their findings by mailing a report to every resident in the watershed. They knew that reversing the salmon decline would require mobilizing local landowners who had deep place attachments with the watershed they called home.

During this same period, ten tribes formed the InterTribal Sinkyone Wilderness Council, which aimed to halt logging of Sinkyone Wilderness State Park's remnant old-growth rain forest, to protect declining salmon, to promote healing, and to revitalize the tribes' cultural relationships with their ancestral lands and waters. On the 3,844 acres they acquired from the Trust for Public Land, the council completed numerous salmon habitat and watershed restoration projects, and later, when the tribes were given a crucial 164 acres at the "Four Corners" region of the watershed, they reconnected with the very spring from which the Mattole River originates. These efforts underscore the profound interdependence within the ecosystem: The salmon rely on the river, the river depends on the trees, and the trees draw sustenance from the salmon and the watershed. In turn, the people depend on the health and vitality of all three, illustrating the deep and enduring connections between the ecosystem and the cultural heritage of the tribes.

State and federal agencies, including the BLM, the California Department of Forestry and Fire Protection (Cal Fire), and the California State Coastal Conservancy, soon noticed the herculean efforts of the nonprofits and tribes. Key partnerships were established, and funding from these agencies fueled progress. Old logging roads were decommissioned and regraded, prescribed burns were reintroduced on both private and public land, hundreds of thousands of trees were planted, and innovative coastal prairie projects were initiated near the Mattole estuary.

For almost half a century, these grassroots organizations and tribes have tirelessly worked along every portion of the sixty-two-mile waterway that runs from the spring feeding the first trickle of the Mattole River to its mouth meeting the Pacific. The collaborative efforts have not always been easy, and the mission to save the watershed and the salmon has been marked by challenging ideological conflicts and the complex realities of anthropogenic environmental change. It has often been two steps forward, one step back, and certain initiatives have worked better than others. Some residents have resisted participation while others have embraced it. Drought and flood continue to threaten the work that's been done, and yet, the sustained journey has ultimately yielded progress for the salmon.

From a low point in 1990, when only two hundred returning salmon were counted, today the numbers consistently reach almost a thousand, and even though these totals are still far below the original estimates, the recovery efforts signal a hopeful trend. The watershed as a whole now boasts burgeoning forests, increased streambed complexity, and reduced sedimentation, marking a slow yet undeniable march toward ecological balance.

Freeman House reflected on the decades of effort in his poignant memoir *Totem Salmon* (1999). Capturing the reciprocity between the community and the natural world, he wrote, "We are bound to each other through our tentative and cautious engagement with the very processes of creation. . . . Most importantly, we have begun our engagement with a place, a place defined by the waters of the river we work in, a place where we may yet come to be at home." For House, for local residents, and for the tribes who have devoted their lives to revitalizing their land and laboring for the survival of a species, the gift they received in this reciprocal exchange was the gift of *belonging*—to a watershed, to a species, and to one another.

This sense of belonging and interconnectedness challenges common perceptions of wildness. So much of the language of wildness and getting back to nature is about searching for a sense of freedom, a release from the burdens of routine daily life. Much like in my longing for the outdoors that preceded my trip to the Trona Pinnacles, I too sought this escape. Yet, in seeking this liberation, there can be a loosening of ties and a loss of obligation. House's reflections prompt us to think beyond the individual and to recognize our collective role in the responsibility of care that accompanies the right of access to public lands. Perhaps true freedom is found not in the absence of ties but in our attachment with place and the stewardship we share with the land and each other.

On our final day in the King Range National Conservation Area, Noah and I opted for a grand finale hike. Our route would ascend the hills overlooking the lagoon, cross the coastal prairie, switchback down to the historic Punta Gorda Lighthouse, and then return to the campground along the beach via the Lost Coast Trail.

High up on the ocean cliffs, six turkey vultures welcomed us with outstretched wings held perfectly still. They perched stoically, side by side, like statues. Walking past these strange creatures felt like a rite of passage. As we continued climbing the steep gravel road toward Strawberry Rock, we found ourselves in a liminal space, caught between contrasting worlds. To our west were the aqua- and midnight-blue waters of the Pacific, the bright sun a spotlight on the vibrant ocean. To the south, where our path continued, a dense fog had settled over the bluff, all but blinding our view ahead. And to the north, some seven hundred feet below, lay the last few miles of the Mattole River.

From our elevated vantage point, the river and lagoon were on glorious display. The scene left me in a state of gratitude for all those who have worked to protect these lands and the river running through them: for Clement Miller, who introduced the very idea of a Conservation Area, and for Donald Clausen, who saw it through; for Joe Duncan and the Native Mattole people, who laid the foundational example of reciprocity; and for all the local advocates and tribal members who have spent their days engaging with the watershed and the salmon that depend on it.

Come autumn, heavy clouds will roll in off the sea and dump rainstorm after rainstorm on the King Range. The moisture will nourish the soil and resuscitate dormant flora. Forgotten creeks will start up again and tumble down the precarious western slope. Fog will linger like a winter blanket, shrouding lower elevations in mist. And gathering on either side of the sandbar, waiting patiently for the rains to open up passage, will be the salmon. For those in the lagoon, the great ocean awaits. For those in the sea, a homecoming.

Heeding the wisdom of the Indigenous Mattole, who cautioned against staring at the water for too long, we turn our gaze away and head south, straight into the fog.

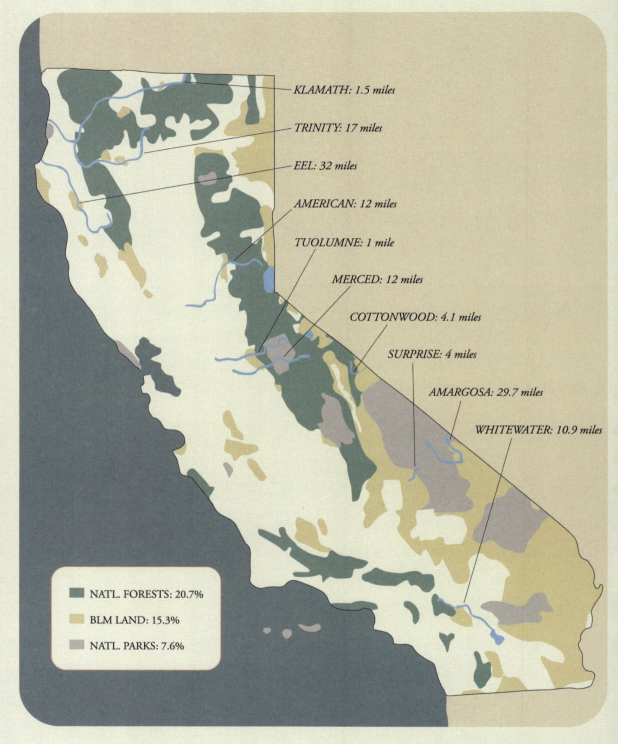

# WILDERNESS, REIMAGINED:
## *A Walk in the River*

*Forty-two months* after my first trip to the Mojave, I stood again in front of the 1989 California BLM wall map that had started this project. Returning to it felt like visiting an old friend. Over thirty-one trips and ninety-six days in the field, I had become intimately acquainted with nearly all of the yellow-shaded BLM lands on the map, and each landscape had endeared itself to me. The way I learned to move in pilgrimage in the Mojave had become the foundation for all my subsequent trips. Taken together, I had walked 381 miles across the deserts, mountains, and prairies of BLM lands. I felt I had, by all accounts, taken a good look around the Golden State.

There was only one unvisited locale left on my list, and I had been saving it for last. Tucked away in a remote enclave of northwestern California, framed in by Highway 101 in the east and the crumbling cliffs of the Pacific in the west, is the Elkhorn Ridge Wilderness.

Spanning 11,112 acres, the Elkhorn Ridge Wilderness is ranked sixty-ninth in size among California's seventy-eight BLM-managed Wilderness Areas. The South Fork of the Eel River flows directly through the Wilderness, carving a steep canyon over several miles. In 1981, thirty-two miles of the river were federally designated as Wild and Scenic, making it one of ten rivers in California managed by the BLM under this designation.

Accessing the Elkhorn was proving difficult because of its modest size and the fact that it was nearly landlocked by private properties. The only public road that managed to get close to the Wilderness was unmaintained and, generally, in disrepair due to downed trees and landslides caused by wildfires and heavy rains. Phone calls to the local BLM field office yielded very little information about the Wilderness, and other avenues I had learned to lean on—the BLM website, conservation

Trinity Wild and Scenic River

Whitewater Wild and Scenic River

groups, hiking apps, and blogs—also turned up empty. Not only had I not found a way in, I also hadn't found anyone who had actually been there. I wondered if a trip was even possible.

Despite the challenge, however, the idea of immersing myself in a federally designated Wilderness Area persistently occupied my thoughts. Such a rugged, deliberately remote location seemed like the perfect place not only to conclude my journey but also to grapple with a question that had been increasingly tugging on my mind: Is a pristine wilderness somehow more valuable than other natural areas? In his seminal essay "The Trouble with Wilderness," environmental historian William Cronon poses a challenge: "To protect the nature that is all around us, we must think long and hard about the nature we carry inside our heads." The Elkhorn Ridge Wilderness seemed like a fitting place to wrestle with sentiments I had long held sacred.

A breakthrough finally came when I closely examined online maps and discovered that the southern edge of the Wilderness borders the Angelo Coast Range Reserve, an 8,051-acre natural laboratory managed by the University of California and dedicated to field research. The reserve's manager told me on the phone that late fall was the only viable time for entry into the Wilderness, when the river's depth would be low enough after the dry summer weather and before the winter rains.

He kindly gave me permission to park overnight in the reserve's day-use lot and then emailed me the gift I needed most: a hand-drawn map. The sketch featured lines, dots, and a few notable landmarks scribbled in cursive. He insisted it wasn't going to be easy. Just getting to the edge of the Wilderness boundary required a seven-mile hike through the reserve along gravel roads and narrow trails. And once we finally arrived to the Elkhorn, he noted, there were no trails, bridges, signage, or infrastructure of any kind. Making things even more difficult, the canyon terrain was so steep the only way to experience the Wilderness was by walking in the river itself. "Hopefully the rains don't come before you do," he said candidly before we hung up. From what he described, it seemed that this Wilderness Area was existing in the kind of isolation originally intended by the Wilderness Act of 1964.

Surprise Canyon Creek Wild and Scenic River

In the aftermath of World War II, America grappled with a growing environmental crisis. Unregulated manufacturing industries and big agriculture choked lakes, rivers, soil, and air with harmful emissions and chemicals. Suburban sprawl led to a surge in demand for wood, which led to extensive logging in previously untouched old-growth forests of the Pacific Northwest and by companies using unsustainable methods that resulted in widespread clear-cutting and significant ecological damage. In response, a conservation ethic matured across the political spectrum, spurred by these obvious signs of environmental decline. Out of this deepening crisis, lawmakers passed the Wilderness Act as a call for renewal, a legislative effort to preserve the nation's most primitive landscapes.

The passage of the act, first introduced by Republican representative John Saylor and Democratic senator Hubert Humphrey, was fueled by the fervent advocacy of conservation luminaries. Among them, naturalist Margaret Murie, dubbed the "grandmother of the conservation movement" by the Wilderness Society, staunchly spoke on behalf of the act, and Rachel Carson roused the public with her book *Silent Spring* (1962), which sold two million copies in its first two years and catalyzed a firestorm of support for environmental action.

Another prominent voice of support came from Wallace Stegner's famous "Wilderness Letter" (1960), which articulated the existential need for wilderness, framing it as a cornerstone of American identity and sanity. His captivating words of warning reverberated through the halls of power and resonated in the hearts and minds of an entire country: "Something will have gone out of us as a people if we ever let the remaining wilderness be destroyed; if we permit the last virgin forests to be turned into comic books and plastic cigarette cases; if we drive the few remaining members of the wild species into zoos or to extinction; if we pollute the last clear air and dirty the last clean streams and push our paved roads through the last of the silence . . ."

The linchpin to the act's success was the steadfast and resolute Howard Zahniser, of the Wilderness Society, who painstakingly penned the bill. Between 1956 and 1964, the bill went through an astonishing sixty-six drafts, eighteen congressional hearings, and sixteen thousand pages worth of testimony. Zahniser's wife, Alice, once remarked on his tenacity, "He'd just go and go, often thirty hours at a stretch

Cottonwood Creek Wild and Scenic River

without sleeping." When compromise and consensus were finally achieved, the bill passed almost unanimously in the house, 374–1, and it was signed into law by President Lyndon Johnson on September 3, 1964. "If future generations are to remember us with gratitude rather than contempt," Johnson declared, "we must leave them more than the miracles of technology. We must leave them a glimpse of the world as it was in the beginning, not just after we got through with it."

Passage of the Act was downright historic. In doing so, the United States became the first nation in the world to define and protect wilderness areas through law. Practically speaking, the law set in motion a number of outcomes: One, it established the National Wilderness Preservation System, which at the outset protected 9.1 million acres of Wilderness and today covers nearly 112 million acres. Two, the act provides the highest standard for public land preservation that exists in the country, safeguarding protected spaces from development, resource extraction, and mechanized transport while emphasizing the value of Wilderness Areas as vital habitats where plants and animals take priority over profits and human ambitions. And three, the act provides a carefully articulated definition for Wilderness: "A wilderness, in contrast with those areas where man and his

Mojave River

works dominate the landscape, is hereby recognized as an area where the earth and its community of life are untrammeled by man, where man himself is a visitor who does not remain."

---

The idea of Wilderness Areas being "untrammeled by man" arouses one of wilderness's most complicated paradoxes. On the one hand, the act protected ecosystems (mostly in the West) from facing the demise other areas (mostly in the East) had already suffered by way of development and extraction that left the landscape fragmented into near oblivion. But the flip side was the way the bill further marginalized Indigenous populations who had lived on and cultivated the land sustainably for thousands of years. The contradiction in the bill's language "untrammeled by man" is not lost on North America's first peoples.

Yet, in spite of the Wilderness Act's shortsighted viewpoints, tribes today are generally in favor of Wilderness designations for the enduring protection they extend to landscapes, and the designation of the Elkhorn Ridge was no exception. It was part of the 2006 Northern California Coastal Wild Heritage Wilderness Act package, and its passage was a victory for local tribes and conservation groups alike who had been advocating for protection of the ancient forests and vital salmon and steelhead habitats, areas of deep cultural significance and ecological importance.

In her excellent book on Native California knowledge and practices, *Tending the Wild*, ecologist M. Kat Anderson dispels the myth of the uninhabited wilderness: "Every day of every year for millennia, the indigenous people of California interacted with the native plants and animals that surrounded them. They transformed roots, berries, shoots, bones, shells, and feathers into medicines, meals, bows, and baskets and achieved an intimacy with nature unmatched by the modern-day wilderness guide, trained field botanist, or applied ecologist. . . . Through twelve thousand or more years of existence in what is now California, humans knit themselves to nature through their vast knowledge base and practical experience."

Her poetic words "knit themselves to nature" serve as an important reminder. Before colonialism, man wasn't a "visitor who does not remain" but a caregiver who remained indefinitely. This historical symbiosis challenges the vision articulated by President Johnson when he spoke of leaving "a glimpse of the world as it was in the beginning." The reality, however, is that the "beginning" he imagined was, in fact, a world already intricately shaped by thousands of years of human imprint.

The Elkhorn Ridge Wilderness exists within the greater Eel River watershed, which includes the South, Middle, and North Forks of the Eel River and is characterized by the rugged Coast Ranges that run parallel with the Pacific Ocean. It remains the ancestral land of the Cahto, Sinkyone, and Yuki Indigenous peoples, who have inhabited this part of northwestern California for thousands of years and maintain communities there today.

The Yuki people's history of near removal from the land was meticulously documented by the ethnographer and anthropologist George M. Foster in his work *A Summary of Yuki Culture*, from 1944. Foster offered a sobering summary of the injustices faced by the Yuki in the middle of the nineteenth century, when they were nearly exterminated by settlers and a vigilante group called the Eel River Rangers: "By the time the Eel River Rangers disbanded in 1860, evidence suggests [the region] had been totally depopulated of Yuki people. Conservatively, approximately six hundred Native Americans were directly killed . . . and many hundreds more taken prisoner and forced into slavery."

The Yuki themselves were a collective of several subgroups united by a shared language and cultural practices, and among the groups was the Wit?ukomno?m. One of the crucial informants for Foster's work on Wit?ukomno?m culture and language was Eben Tillotson, born to a Wit?ukomno?m mother and a white father after the decade of genocide that scarred the northwest of California. Tillotson spoke often with Foster, offering invaluable insights into the Wit?ukomno?m people's enduring heritage. Foster writes, "Tillotson was a very conscientious informant, and took a real interest in 'setting the Indian word down right.' He told only of what he was sure, admitting that there was much he could not answer." In *A Summary of Yuki Culture*, Tillotson is referenced fifty-two times, significantly shaping the documented record of the Yuki people's language and culture.

Tillotson's legacy of preservation lives on today in his great-great-grandson Joseph Byron. Joseph is a thirty-nine-year-old Wit?ukomno?m who has dedicated his life to revitalizing the very language and culture that Eben had worked to preserve. I had the privilege of interviewing Joseph, and through our conversation I gained a deeper understanding of the complex history of Northern California and of his passionate efforts to weave together the historical and contemporary narratives of his people.

San Joaquin River

Joseph has a youthful face and a tender smile, and he speaks with a forthrightness and surety that seems to mirror qualities his great-great-grandfather had. He shares the experiences of his ancestors not as distant memories but as events that happened recently; "It wasn't that long ago for us," he points out. He holds back tears as he describes the connection between Eben's life and his own, bringing the past vividly into the present and expressing the gratitude he feels for Eben recording their language. Already fluent in two Indigenous languages and on his way to learning a third, Joseph considers it his paramount duty to revive these languages within his community.

In discussing the Wilderness Act's impact, Joseph offers a unique perspective on his homeland, now fragmented by modern land designations. "When I go through the different homelands," he says, "I'll shift into the language that was spoken in that area. We believe our languages are embedded in the landscape and embedded in the waters. And it's our language that will bring healing to the land."

His perspective sheds further light on the conventional dichotomy between wilderness and inhabited areas, and he proposes understanding it instead as a deep, intertwining relationship that is linguistic, spiritual, and familiar. He imagines a reality where the connections among people, their surroundings, and all living beings are inseparable and nurtured by the ancient languages native to these lands. As Joseph explains, "There is an invisible umbilical cord that is attached to us and the lands that we come from, our homelands and our home waters. Even though physically we were removed, spiritually we are still here. As long as the earth and these lands are present, we are still connected to them. And that includes our plant relatives and the animals that are indigenous to these lands and these waters."

---

For my trip to the Elkhorn, I invited my father along, partly because I wanted to introduce him to BLM lands, but also because I knew such remote expeditions are best undertaken with company.

I picked him up at the airport the night before we set off. He wore his familiar, unwavering smile and, true to form, his backpack was already packed for the backcountry. We shared a hug, our first in ten months, and I immediately expressed my excitement and concerns about the Elkhorn, admitting I had almost no clue what we were getting ourselves into. My dad, a veteran of adventurous undertakings, was not rattled in the least. This wasn't the first time we had ventured into relatively unknown territory. Twenty-eight years prior, in the summer of 1995, my dad and I hitchhiked from our home in Michigan to New York City, visiting baseball stadiums along the way. For our final ride into Manhattan, we sat in the back of a pick-up truck, on opposite sides of a Harley-Davidson. Construction workers on the George Washington Bridge waved and shouted as we rode by.

Even then, I have no memories of my dad being worried or fearful in unfamiliar situations. And he certainly wasn't anxious about "A Walk in the River," as I was calling our trip, a nod to one of our shared favorite books by Bill Bryson, *A Walk in the Woods*. We left for the Elkhorn early the next morning, long before the sun appeared.

Ten hours later we found ourselves riding in the back of another truck, this time bouncing through a dense conifer forest in an aging Tacoma driven by the UC Angelo Reserve's manager. He had met us at the day-use parking lot and offered a ride that would shave one mile off our seven-mile trek to the Wilderness boundary. I snapped a blurry photo of my dad and myself as we laughed nostalgically about our previous ride together in the back of a truck. Almost thirty years later, we had traded New York City for old-growth redwoods, the Hudson River for the South Fork of the Eel, and construction workers for bears.

We jumped out of the truck, profusely thanked the manager, and started the six-mile hike along the gravel back roads and trails of the reserve, my hand-scribbled map guiding the way. As we walked deeper into the cover of woods, the shade was a welcome respite from autumn temperatures still hovering close to 80 degrees. Near the end of the map, as if perfectly on cue, the trail came to a sudden halt at a cliff with a panoramic view of the South Fork of the Eel River. We had made it to the very edge of the Wilderness, and the end of the trail meant we were on our own.

From the pointed end of the cliff, we picked the safest route toward the river and began descending a steep slope, half sliding and half falling, our trekking poles the only thing keeping us somewhat upright. We navigated our way through a dense understory of chaparral until we arrived at the river's edge, where the water was shallow enough to rock hop our way downstream. A hundred feet later, near the confluence with a shallow creek, the Eel suddenly drops, narrows, and runs haphazardly through a thicket of willow and boulders, making a maddening scramble through a labyrinth of flora and rock. Even with the light flow of the river, the water shot and spit and foamed its way through the maze with wild unpredictability. I kept an eye out for my dad, sixty-five years young, but he was maneuvering through the obstacle course with relative ease.

We arrived to the official Wilderness Area just as the last shred of sunlight was visible on the ridge high above the river. The scene was exhilarating. The gentle curves in the river were just enough to provide a hypnotic blend of whitewater and slow-moving pools. Douglas firs and cedars stood along the ridgeline and protruded into the darkening sky with gusto. Nestled along the banks, the elder oaks, madrones, and alders cast their reflections upon the mirror-like surface of the pools. Their branches stretched out over the river in gratitude, as if offering a silent acknowledgment for the nourishing moisture.

As the river wound further downstream, we found a pebbled beach replete with enough dried-out driftwood to build a fire and maybe even fashion a proper shelter if things went awry. In spring of that year, when atmospheric rivers had stormed through and turned the Golden State green, the beach must have been completely under water, but after a dry, hot summer, the rushing river had subsided enough to reveal the inviting shoreline. We immediately dropped our packs without deliberating. This was our camp for the night.

I had counted eleven piles of bear droppings on the way in, and I found two more large piles on the beach, one of which looked as if it had only recently run out of steam. The other, completely dried up, looked like it could have kindling potential. I poked sticks into the scat, making them stand up like birthday candles, and then built a campfire structure around it with pieces of silver driftwood.

We pitched our tent, collected more firewood, and snacked our way through enough calories to make up for our breakneck hiking pace. As night enveloped our little corner of the universe, the fire grew steadily, sending flames of light flickering across the river, projecting a wondrous show on the hillside forest. Exhausted, we stretched out on either side of the blaze to properly take in the dueling symphony of water and fire. "This sound never gets old," said my dad. Above the ridge, stars began making their appearance. At the end of a long day, there were few words left between us.

Falling temperatures eventually pushed us toward the retreat of our tent. But just as we squeezed into our sleeping bags, there was a freakishly loud splash on the water. It sounded to me as if a bear had climbed one of the towering trees and

*ELKHORN RIDGE WILDERNESS*
*South Fork Eel River*

South Fork Eel Wild and Scenic River

cannonballed off the top limb into the river. I jumped out of the tent and raised my headlamp toward the sound. Mysteriously, I found nothing.

Despite my dad's best wishes for a good night of rest, I knew there was little chance I'd be sleeping much. There were two things weighing on my mind. One, of course, was the anxiety of waking up to a bear pawing at the tent lining, breathing its muffled sounds of curiosity. I thought through a number of responses in case this happened, the odds of which I estimated at 99 percent.

The second was the Wilderness I had found myself in. For most of my life, I had understood nature as being "out there," in landscapes that were distinctly separate and uniquely other from the urban environments I called home. In the culturally sculpted hierarchy of nature I subscribed to, the most authentic form of wilderness was found in designated Wilderness Areas like this one. For me, these remote landscapes not only contained the most pristine and majestic beauty, they also provided a setting where humans and modern conveniences were almost entirely absent. It was here, I believed, where nature existed in its most elemental form, where I could find my very own Eden.

In "The Trouble with Wilderness," Cronon explains the philosophy I had long held, one inherited from Stegner and many other ardent conservationists: "For many Americans, wilderness stands as the last remaining place where civilization, that all too human disease, has not fully infected the earth. It is an island in the polluted sea of urban-industrial modernity, the one place we can turn for escape from our own too-muchness." These sentiments had mirrored my own; finding my island of nature always meant escaping the city.

But by embracing this philosophy, I had fallen into a mindset that Cronon cautions against: "Idealizing a distant wilderness too often means not idealizing the environment in which we actually live, the landscape that for better or worse we call home." As I lay in the cozy confines of my sleeping bag, it was this dilemma that kept me awake. In making an idol out of the nature found in wilderness, I wondered if I had too often overlooked and devalued the nature found in my everyday surroundings.

Thanks to dozens of pilgrimages through BLM landscapes, each of which had revealed the nuanced beauty found on lands once considered "leftover," my understanding and perspective on nature had been undergoing a monumental shift. Yet, as I pondered Cronon's cautionary sentiments, I had a sneaking suspicion that my evolution hadn't gone far enough.

AMERICAN BLACK BEAR
*Ursus Americanus*

Eventually I fell asleep, ten different times, and I can report that no bears approached the tent (that I know of). It is hard to describe the momentous feeling it is to be greeted by morning after a long night of restlessness in the middle of bear country.

"We made it," I said to my dad with relief. "That last run of sleep was the best I had all night."

"Me too," he replied sheepishly. "I think it was the sheer exhaustion and because my whole body is currently numb."

We shared a long laugh together over my exaggerated fear of bears and his night of discomfort. Hot coffee and the rising sun provided warmth as we headed downriver, deeper into the Wilderness, using the South Fork of the Eel to dictate our path. The most visible signs of autumn were displayed in the radiant, overgrown patches of poison oak, which ran up from the river and painted the slopes red. The oak served as a fiery contrast to the fluorescent greens lighting up the sedge and willows interspersed along the banks. Colossal boulders were stacked and scattered along the river, each one showcasing a horizontal demarcation line of color that revealed the high-water mark of spring floods.

Weathered stones, polished by centuries of flowing water, were strewn from shore to shore, jutting up like miniature islands, creating a choose-your-own-adventure kind of landscape. We hopped and stepped our way downstream and soon found ourselves on opposite shorelines, taking in the wilderness in our own ways. By mid-morning, I was shedding layers and standing barefoot in a shallow stream, admiring darting minnows and taking in the soft yellow flowers of a western goldenrod while my feet grew numb in the cold river. In a moment of communion, I scooped up a handful of water and splashed it against my face.

Crisscrossing the riverbank were game trails etched by the passage of bears, lions, and bobcats, and no doubt traveled by wildlife farther down the food chain. I ventured up a well-trodden path into a mixed evergreen forest where tanoak, canyon live oak, and Pacific madrone flourished. The woodland floor beneath my feet was so thick with leaves, branches, moss, and grasses that it felt like walking on a trampoline. One particular gnarled madrone with low-hanging limbs caught my attention, and I carefully climbed the twisted, peeling trunk until I was two stories up, hidden in a dense canopy of foliage. In the suspended state between earth and sky, the perspective was simultaneously unnerving and exhilarating. Drawing in a deep breath, I closed my eyes, surrendering to the symphony of swaying leaves and cascading water, remembering that gratitude is only limited by our awareness of it.

By the time we reconvened downriver, the sun had reached its zenith in the midday sky. We perched ourselves atop a boulder to get a better view of the river ahead and saw the end of our journey. With the canyon narrowing, the mountains growing steeper, and the river deepening, we had ventured as far as the terrain would permit. I voiced the inevitable: "Looks like this is the end." To commemorate the bittersweet moment, I snapped a photo of my dad standing atop the boulder, as content as ever in nature, his wide smile brimming with joy.

Slowly, we retraced our steps upriver to our temporary beach camp, where we gathered our gear and erased any trace of our transient presence. We wove back through the thicket of willows, ascended the cliff overlooking the river, and cast one final gaze on the landscape we'd likely never witness again.

---

In the days since returning from the Elkhorn, as I've tried to process the wilderness binary that Cronon cautions against, I kept returning to Joseph Byron's poignant and evocative analogy of the invisible umbilical cord. Like children physically connected to their mothers, like Joseph spiritually connected to his ancestral lands, we are all connected to what Cronon calls "the landscape that for better or worse we call home"–the places we spend the majority of our days actually living.

Coming home, I resumed my daily walks around our busy neighborhood in the heart of Los Angeles. My usual three-mile loop runs past an array of houses, apartment blocks, and the pulsating hum of the city, from restaurants and bars to mechanic shops and bodegas. Besides the familiar hellos exchanged with neighborhood regulars, each day brings the possibility of encountering unexpected natural delights and surprises. One morning, a great horned owl made an appearance, standing vigil atop a deodar cedar. On three occasions, a lone coyote has sauntered along the sidewalk just ahead of my footsteps, pausing every few houses to lock eyes with me, stirring memories of the coyote I had encountered in the Amargosa. The ravens are always present, as are a flock of exotically colored parrots, who shriek with enough collective ferocity that even the hard-nosed ravens retreat.

The real gift of the walk, though, is the trees. They line the sidewalks and alleys, fill in yards, and separate one tiny lot from another. My youngest daughter, Vivian, often accompanies me, and together we have been slowly learning their names by studying their bark, leaves, and distinctive features. We pass camphor, jacaranda, strawberry, ash, live oak, southern magnolia, cypress, and Buddhist pine, among dozens of others. The leaves of the pepper and eucalyptus trees offer their generosity through aroma, and we crumble them in our hands, raising their fragrant scent to our noses. Her favorite tree is the purple-flowered jacaranda, which drapes the sidewalks in lavender just as the school year is coming to an end, a sure sign that the freedom of summer is just around the corner.

My favorite tree along the path is one particular western sycamore. Its gnarled roots give way to a richly textured trunk, which is dark and bumpy at the base but becomes smoother and fresher higher up, with bark that features alternating patches of gray and tan. It stands just a few steps off York Boulevard, the main thoroughfare running through the neighborhood. If it had eyes, the views wouldn't be much to write home about: a grocery store to the south, a body shop to the north, a coin-operated car wash on one corner, and a fenced-in lot full of old cars on another. At night, the lights of a taco truck flicker on.

Despite facing an uphill battle for existence here in this urban landscape, the sycamore is somehow thriving. Its trunk, marked with nails and knife engravings, is literally buried in concrete, and it is constantly exposed to exhaust from passing vehicles. Yet every spring, I watch little buds suddenly appear and then morph into magnificent fuzzy green leaves, as if the tree has convinced itself that it's rooted beside a creek. Not long ago, I was perched in the second story of a madrone, in the middle of a wilderness as floristically diverse as any place in California; now, I stood before the sycamore, alive and prospering in an environment many would consider the antithesis of wilderness.

I had walked by this sycamore hundreds of times without ever actually *seeing* it. I hadn't looked beyond the concrete, parked cars, street signs, and noise that define my neighborhood to notice the diversity of trees, the poppies growing out of sidewalk cracks, the hummingbirds diving in and out of nectar-rich succulents, or the lizards scurrying between fence lines. The poet Gary Snyder once declared that "a person with a clear heart and open mind can experience the wilderness anywhere on earth. It is a quality of one's own consciousness."

It took four years of walking BLM landscapes to slowly rearrange the nature I carried inside my head, and it took that final trip to the Elkhorn Ridge Wilderness to finally—with the help of Cronon's prompting—bring everything I had learned back home. Each destination, each encounter, has been a chapter in a larger story of awareness.

The sycamore, standing resiliently amidst the urban chaos, serves as a vivid reminder that beauty, wildness, and nature are not confined to remote wilderness but are woven into the very fabric of our everyday lives.

# EPILOGUE

*Nine months after* exploring the Elkhorn Ridge Wilderness, and a decade after my inaugural trip to the Trona Pinnacles, Kari, the kids, and I find ourselves in the northern reaches of the King Range National Conservation Area for a late-July camping trip. Our kids, now fifteen, twelve, and nine, have become seasoned veterans at navigating BLM lands, as familiar with their landscapes as they are with their histories. We are joined by some family and friends who all made the long trip north for one last summer adventure before the alarm clock rings in another school year.

Near the Mattole Beach Campground, a potholed gravel road leads away from the beach, climbing seven hundred feet until the topography flattens out along a bluff overlooking the Pacific. Our dusty cars follow the steep switchbacks upward for a few miles—the same road Noah and I had walked up three years prior—until the coastal prairie transitions into a dense forest of moss-covered Douglas firs. We park along the ecotone where tall grasses meet the verdant green of the trees, and we begin the mile hike out to Windy Point, aptly named for the strong gales that sweep in from the northwest. The hike starts along a two-track dirt road and then picks up a barely visible path through knee-high grass, golden in July.

The kids lead the way, scattered along the path, alternating between running and walking, fueled by candy and the pull of the ocean. The adults trail behind, conversing and taking in the majesty of the bluff, some for the very first time. A thin fog hovers over the landscape, moving southeast in between bursts of sun rays, and a pair of distant cypress trees are silhouetted against the sky, their rigid branches all pointing east from the persistent winds. I hand my camera to our oldest daughter, Stella, who confidently slips the strap around her neck and glides ahead, ready to capture the landscape in her own unique way.

The setting is just as spectacular as I remember it. My eyes move between the white-capped waves colliding out at sea and the steeply pitched mountains that

Coastal prairie

rise up against it. But for all the natural beauty that exists, the imprint of humanity is everywhere.

The bluff is dotted with cow patties, which have been scattered by the hoofed creatures permitted to graze here. Stones are stacked and placed in circles, remnants of fire rings used by dispersed campers who come for hiking, hunting, and solitude. Nearby is a fifteen-acre field of wildflowers and native grasses that were painstakingly seeded and planted by members of and volunteers from the Mattole Restoration Council.

A hand-painted sign made by local elementary school children offers a reminder to visitors: "Protect our prairies," it says in faded letters. Evidence of large burn scars are visible up and down the bluff, where encroaching Douglas firs were piled and burned in an ongoing restoration effort to bring back the coastal prairie. And down on the narrow beach below, hikers trudge with heavy backpacks along the Lost Coast Trail.

It is a scene that perfectly illustrates the BLM's multiple-use mandate. Even on this relatively small patch of coastal prairie, grazing, recreation, and restoration projects have left a lasting impression. Returning here once again, I'm reminded of Freeman House's declaration from *Totem Salmon* about the work the Mattole Salmon Group was doing in this very locale: "We have begun our engagement with a place," he wrote. After a decade of researching and walking these lands, I believe it's this opportunity for engagement that compels us to shift our relationship with the land from what we can take to one that asks what we can offer.

When visiting our beloved National Parks can sometimes feel like entering a museum, complete with parking lots, gift shops, and long lines, nature can often feel as if it's behind a velvet rope. Conservation in these areas, as well as in Wilderness Areas, follows a "fortress" model, which aims to protect nature by creating spaces isolated from human impact. While there is certainly a place for this model, and while I'm grateful for our National Parks and all the wonderful experiences they provide, this limited ideal often means engagement with nature is restricted to observation. We visit, we admire, and then we depart.

BLM lands, in contrast, summon a public land ethic that requires not just a more nuanced appreciation of beauty but a transformative perspective on our relationship with these lands. Instead of seeing them as inferior, we can recognize these landscapes as full of great beauty and biodiversity, home to thousands of species that rely on them. Rather than viewing them as impaired or beyond worth, we can see them as places where meaningful work can be done, where the opportunity to recreate comes with an obligation to steward and care for these places in return, visiting them not as playgrounds or treasure boxes to plunder but as extensions of home.

Perhaps through this relationship of reciprocity—what David Abram called "the ceaseless give and take, the flow that moves in two directions"—we can not only deepen our attachment but find our own sense of belonging.

With the odds stacked against these lands, they have endured. They are essential not merely for ourselves but for the embrace they offer to the animal and plant kingdoms, for the sanctuaries of habitat and biodiversity they provide, for the ways they alleviate the impacts of a warming planet. In a world besieged by dissension, these lands can become refuges of sanity and solace, where present and future generations can find common ground, where our kids can feel the earth and watch the moon rise, and then have some stories to tell when they get home.

---

As we stand together out at Windy Point, conversation momentarily comes to a halt. We gather along the bluff and gape at the waves that crash on the black sand beach and rocky islets some five hundred feet below. Sea lions bark in the distance, their guttural roars the only sound cutting through occasional gusts of wind. The emerald-colored waters along the shoreline eventually give way to a deep navy before melting into the light-blue line of the horizon, while mountains to the north and south stack on top of one another. While we collectively take in the scene, I turn around and face the continent.

My mind spins through the experiences I've collected over the past decade of traveling these lands. The freezing winter sunrise hikes, the lonely nights beside a fire, walking up and down mountains, pilgrimages across the Mojave, and all the laughter shared with traveling companions. So many moments of awe, so many miles under my feet, so many rocks in my shoes and dust on my arms and sweat under my hat.

I take a deep breath of gratitude for all that these lands have given my family and me, and for the ways they have challenged our relationship with nature, pushing and prodding us to *involve* ourselves more deeply in what William Cronon called "the great task of struggling to live rightly in the world—not just in the garden, not just in the wilderness, but in the home that encompasses them both."

Looking back at the golden prairie, I feel a surge of hope and anticipation as the journey continues. We have begun our engagement with place.

Windy Point

# ACKNOWLEDGMENTS

*Writing this book* has required the most sustaining amount of fortitude and patience I have experienced in my professional life. I shudder to think where I would be without the incredible individuals who have profoundly influenced its creation. To each of these fine humans, I am forever indebted.

To Justin, who first introduced me to BLM land a decade ago.

To my traveling companions who traversed these landscapes with love and humor, and who offered many keen observations along the way. Paul, Sam, Noah, Asher, Julie, Joni, Aaron, Jon, and my father, Dennis—thank you for walking with me.

To Rebekah Nolan, whose maps and illustrations are as exquisite as they are inspiring, and to her partner, Eric, whose collaboration has brought immense creativity and joy to this project. And to Noah Smith, designer extraordinaire, for his timeless cover.

To all the civil servants I've met at the BLM, especially Leisyka Parrott, the first field manager who agreed to an interview before I had any idea what I was getting myself into, and Jesse Pluim, for kindly answering a thousand questions along the way.

To everyone at the Conservation Lands Foundation and their grassroots network of friends organizations, whose unwavering commitment to protecting, defending, and expanding the future of BLM lands has been a constant source of inspiration.

To all those who offered their informative expertise during field visits and interviews: Joseph Byron, Bob Schneider, McKenzie Long, Ryan Henson, Bob Wick, Andrew Fulks, Emma Fryer, Aaron Shintaku, Ryan Haack, Rich Benson, Wendy Schneider, Mark Kenyon, Obi Kaufmann, Jora Fogg, Kayla Browne, Peter Steel, Lynn Boulton, Neil Kornze, Nathan Queener, Hugh McGee, Mason Voehl, and Sandra Schubert.

Thanks to all the books along the way and the writers who wrote them. Taken as a collection, they were my north star.

Many thanks also to the three authors whose works often carried me through stormy days of writer's block and copious amounts of self-doubt. The way reading another's writing somehow helped me overcome what felt like a mountain of creative paralysis can only be described as a miracle, as if the spirit of the writer were speaking to my inner spirit in a secret dialogue. So, my deepest regards to Chris Dombrowski, Gretel Ehrlich, and Robert Macfarlane.

My gratitude to Ashley Harrell for first putting my Forgotten Lands Project on the map and introducing me to Glenn Stout, a prolific author who generously offered sage counsel on my book proposal.

To Marc, who gifted me the BLM wall map that opens the book. To Dennis, who skillfully kept my small furniture business going each time I was away. To Elisa, who confidently gave me the push I needed to get started.

To Marthine Satris, my brilliant and patient editor: I still remember exactly where I was when her life-changing email found my inbox. Her insightful and thoughtful edits enhanced every single page. Thanks also to Lisa K. Marietta for her incomparable copyedit, to Archie Ferguson for directing the beautiful interior layouts, to Kalie Caetano, Megan Posco, and ill nippashi for their far-reaching marketing efforts, and to Emmerich Anklam for holding it all together. And to the rest of the amazing crew at Heyday who had a hand in this lengthy book-making process: Steve Wasserman, Gayle Wattawa, Terria Smith, Chris Carosi, Marlon Rigel, Eve Sheehan, and JiaJing Liu.

To my small but mighty group of inner-circle friends and fellow camping junkies—Paul, Brooke, Sam, Rose, Noah, Audrey, Aaron, and Kress (and all their kids)—my sincere love and gratitude.

To my forever-curious mother, a writer herself, and my adventurous father, both of whose enthusiastic support for this project was a constant source of encouragement. To my sisters—Jamie, Joni, and Julie—whose collective tenderness and dream chasing has made me a better human.

And finally, to my dear family, whose love could sustain me for a thousand loops around the sun. To my wife, Kari, who enthusiastically supported each journey and meticulously read every draft with a discerning eye—you are a guiding light of strength and abiding love. To my long-lost daughter, Margot June, whose spirit kept me company on many freezing sunrise walks. And to our living children—Stella Rose, Leo James, and Vivian Lucia—I love you beyond words (though I'll keep trying). Your endless engagement and rapturous curiosity in many of these landscapes helped me see them with fresh eyes. This book is a love letter to your future.

# GUIDE FOR EXPLORATION

*Should you find yourself* in the enviable position of having some leisure time, whether it be for a weekend or a month, and if these idle hours lead you to the great outdoors, I hope you'll consider a trip to our public lands managed by the BLM.

This guide is meant to be a starting point, a get-the-ball-rolling kind of adventure manual. One of the serendipitous qualities of BLM lands is how they necessitate a more robust set of research tools in order to navigate them successfully. Many of the hurdles I ran into at the beginning of this project—namely, the information gap and the lack of storytelling about these areas—are still present today. This means you'll often have to dig deeper, and for longer, than what is required

Highway 120

Travertine Hot Springs

Pit River Canyon Wilderness Study Area

when you're visiting more traveled locales that have already been covered by a thousand adventure blogs, on which the information has been neatly reduced to bullet points and GPS coordinates.

But there's a great beauty in the research process. Like slowly turning the pages of a favorite novel, the story of the landscape unfolds as you stitch together facts, data, maps, and perhaps some scattered images. You might even gain valuable insights beyond just location and access—knowledge that adds color and vitality to the public lands you'll soon visit. What kind of wildlife and plants exist there? What about the geographical features, the climate, and the history of the place? After this thorough inquiry, you'll find your reward: a landscape waiting for your arrival.

This was often my experience as I explored BLM lands across California. I would start with a landmark, trail, or campground and build my research from there. Some areas, like National Monuments and Recreation Areas, offer more useful information, while other landscapes are almost completely off the radar. If you're a novice in uncharted territory, like I once was, I recommend beginning your exploration on lands with the easiest access. Let these serve as gateways to more isolated landscapes, which often require more gear, knowledge, and patience. I hope the following exploration resources and tips help you on your way.

Crowley Lake Campground

NATIONAL CONSERVATION LANDS: These are the areas in California that are most accessible and often have established recreation infrastructure already in place, such as campgrounds, hiking trails, mountain biking access, hunting areas, and designated OHV use. Many of these Conservation Lands also have helpful brochures that include detailed maps and highlight the history and landmarks. Conservation Lands include National Monuments, Conservation Areas, Scenic Areas, Wild and Scenic Rivers, Wilderness Areas, Wilderness Study Areas, and National Trails. https://www.blm.gov/programs/national-conservation-lands/california

CAMPGROUNDS: At the time of writing, there are sixty-six BLM campgrounds in the Golden State, most of which can be accessed by two-wheel-drive vehicles. All BLM campgrounds (including each of the campgrounds on the travel map that opens this book) are first-come, first-served, and the cost ranges from free to around $10 a night. Most campgrounds have basic amenities, from vault toilets and fire pits to picnic tables and shade structures. However, most of these camps do not have water or trash service, so be prepared! I always travel with a five-gallon water container, toilet paper (restrooms are frequently out), and several heavy-duty trash bags for collecting my own trash and picking up litter around camp. Everything you bring in needs to be packed out.

Bodie Hills

DISPERSED CAMPING: This form of camping, which takes place outside of developed campgrounds, usually along a secondary road pullout or in a clearing, provides a more rugged outdoor experience for self-sufficient campers. Most BLM land is open to this kind of camping, with some exceptions. Here are some useful rules to follow:

- The guiding principles of Leave No Trace are essential for dispersed camping. https://lnt.org/why/7-principles/

- Dispersed campsites are often found on back roads through BLM land and may not be marked. Popular spots are identifiable as flat areas with signs of previous use. Please use existing sites to avoid creating new disturbances.

- I highly recommend calling the local BLM field office (more on that below) to ask about where to find dispersed camping areas in the landscape you'll be visiting.

- If you plan on making a campfire, please call the BLM field office for information on whether it is allowed. You will also need a permit to build a fire while dispersed camping. https://permit.preventwildfiresca.org/

- Dispersed camping is generally allowed for fourteen days within a consecutive period of twenty-eight days.

- You need to be fully self-contained, as there are no amenities available.

- Please place your campsite at least one hundred feet from any stream or other water source.

- DO NOT drive across the landscape in search of a site. There are often very obvious areas just off the road where you can park and set up. Camp on bare soil whenever possible, to avoid damaging or killing plants and grass.

FORGOTTEN LANDS PROJECT: I launched the Forgotten Lands Project to inspire appreciation, engagement, and protection of BLM-managed public lands through storytelling, exploration, and collective advocacy. The site features written and visual dispatches from BLM lands in California and across the western US, along with information on trips I lead and conservation opportunities. It serves as equal parts guidebook, inspiration, and travelogue. forgottenlandsproject.com

Clear Creek Management Area

SAFETY CONCERNS: It is crucial to learn about the specific activities permitted on the lands you are using and how to engage in them safely. The most common safety concerns occur on lands where hiking, equestrian activities, and mountain biking are allowed. And whether you are a hunter or a hiker, be aware that simultaneous use of lands by these groups can be especially dangerous if either party is unaware of the other's presence. Please contact the local BLM field office to learn more.

BLM WEBSITE: The BLM website can sometimes be equal parts helpful and frustrating, but it's always worth checking as a first point of contact. While some areas' pages will offer more information than others, each page will at least provide a phone number for the local BLM field office that is in charge of managing the area or landmark you're researching. https://www.blm.gov/california

San Joaquin River Gorge Special Recreation Management Area

BLM FIELD OFFICES: There are fourteen BLM field offices across the Desert, Central, and Northern Districts within California. These offices manage the land within their jurisdiction, and the BLM employees who work at these offices are generally very helpful and knowledgeable.

- Desert District: https://www.blm.gov/office/california-desert-district-office

- Central District: https://www.blm.gov/office/central-california-district-office

- Northern District: https://www.blm.gov/office/northern-california-district-office

CONSERVATION ORGANIZATIONS: I can't possibly say enough about the tribal communities and conservation organizations and the important efforts they lead; they are the beating heart of engagement with BLM lands. All around California and across the western US, these groups are on the ground working tirelessly to educate and engage their communities. They help protect access, enhance recreation, facilitate restoration projects, and engage in the tedious work around policy and resource management planning that helps shift the imbalance between protected and unprotected BLM lands. I would encourage everyone to find their nearest conservation nonprofit. Among those you might connect with are:

**Alabama Hills Stewardship Group**
**Amargosa Conservancy**
**Backcountry Hunters and Anglers**
**Cache Creek Conservancy**
**California Native Plant Society**
**CalWild**
**Carrizo Plain Conservancy**
**Conservation Alliance**
**Conservation Lands Foundation**
**Council of Mexican Federations in North America**
**Friends of the Carrizo Plain**
**Friends of the Inyo**
**Intertribal Sinkyone Wilderness Council**
**Mattole Restoration Council**
**Mattole Salmon Group**
**Mojave Desert Land Trust**
**Native American Land Conservancy**
**Sanctuary Forest**
**Theodore Roosevelt Conservation Partnership**
**Tuleyome**

Granite Mountain Wilderness

# SELECTED BIBLIOGRAPHY

Abbey, Edward. *Desert Solitaire: A Season in the Wilderness*. New York: Ballantine Books, 1968.

Akins, Damon, and William Bauer Jr. *We Are the Land: A History of Native California*. Oakland: University of California Press, 2022.

Anderson, M. Kat. *Tending the Wild: Native American Knowledge and the Management of California's Natural Resources*. Berkeley: University of California Press, 2013.

Bernard, Ted, and Jora M. Young. *The Ecology of Hope: Communities Collaborate for Sustainability*. Gabriola Island, BC: New Catalyst Books, 1997.

Berry, Wendell. *The Long-Legged House*. New York: Harcourt, Brace and World, 1969.

Brinkley, Douglas. *The Wilderness Warrior: Theodore Roosevelt and the Crusade for America.* New York: HarperCollins, 2009.

———. *Rightful Heritage: Franklin D. Roosevelt and the Land of America*. New York: Harper, 2016.

———. *Silent Spring Revolution: John F. Kennedy, Rachel Carson, Lyndon Johnson, Richard Nixon, and the Great Environmental Awakening*. New York: Harper, 2022.

Carlson, Robin Lee. *The Cold Canyon Fire Journals: Green Shoots and Silver Linings in the Ashes*. Berkeley: Heyday, 2022.

Carson, Rachel. *Silent Spring*. Boston: Houghton Mifflin, 1962.

Connors, Philip. *Fire Season: Field Notes from a Wilderness Lookout*. Chicago: Ecco, 2012.

Cronon, William. *Uncommon Ground: Rethinking the Human Place in Nature*. New York: W.W. Norton, 1995.

Dant, Sara, and Tom Udall. *Losing Eden: An Environmental History of the American West*. Lincoln: University of Nebraska Press, 2023.

Davis, Steven. *In Defense of Public Lands: The Case Against Privatization and Transfer*. Philadelphia: Temple University Press, 2018.

Devoto, Bernard, Patricia Nelson Limerick, and Douglas Brinkley. *The Western Paradox: A Conservation Reader*. New Haven: Yale University Press, 2008.

Dillard, Annie. *Pilgrim at Tinker Creek*. New York: HarperCollins, 2007.

Dombrowski, Chris. *The River You Touch: Making a Life on Moving Water*. Minneapolis: Milkweed Editions, 2022.

Ehrenreich, Ben. *Desert Notebooks: A Road Map for the End of Time*. Berkeley: Counterpoint, 2021.

Ehrlich, Gretel. *The Solace of Open Spaces*. New York: Penguin Books, 1985.

Flores, Dan. *Coyote America: A Natural and Supernatural History*. New York: Basic Books, 2017.

Foltz, Bruce V., and Robert Frodeman. *Rethinking Nature: Essays in Environmental Philosophy*. Bloomington: Indiana University Press, 2004.

Foster, George McClelland. *A Summary of Yuki Culture*. Berkeley: University of California Press, 1944.

Gessner, David. *All the Wild That Remains: Edward Abbey, Wallace Stegner, and the American West*. New York: W.W. Norton, 2016.

———. *Leave It As It Is: A Journey Through Theodore Roosevelt's American Wilderness*. New York: Simon and Schuster, 2021.

Gros, Frédéric. *A Philosophy of Walking*. London: Verso Books, 2014.

Grover, Quinn. *Wilderness of Hope: Fly Fishing and Public Lands in the American West*. Lincoln: University of Nebraska Press, 2019.

Hansman, Heather. *Downriver: Into the Future of Water in the West*. Chicago: University of Chicago Press, 2019.

House, Freeman. *Totem Salmon: Life Lessons from Another Species*. Boston: Beacon Press, 2007.

Kaufmann, Obi. *The Forests of California*. Berkeley: Heyday, 2020.

———. *The Deserts of California*. Berkeley: Heyday, 2023.

Kenyon, Mark D. *That Wild Country: An Epic Journey Through the Past, Present, and Future of America's Public Lands*. New York: Little A, 2019.

Ketcham, Christopher. *This Land: How Cowboys, Capitalism, and Corruption Are Ruining the American West*. New York: Penguin Books, 2020.

Kolbert, Elizabeth. *The Sixth Extinction: An Unnatural History*. New York: Henry Holt, 2014.

Leopold, Aldo. *A Sand County Almanac: With Essays on Conservation from Round River*. New York: Oxford University Press, 1966.

Leshy, John D. *Our Common Ground: A History of America's Public Lands*. New Haven: Yale University Press, 2022.

Long, McKenzie. *This Contested Land: The Storied Past and Uncertain Future of America's National Monuments*. Minneapolis: University of Minnesota Press, 2022.

Louv, Richard. *Last Child in the Woods: Saving Our Children from Nature-Deficit Disorder*. Chapel Hill: Algonquin Books, 2005.

Luong, QT, Sally Jewell, and Ian Shive. *Our National Monuments: America's Hidden Gems*. San Jose: Terra Galleria Press, 2021.

Macfarlane, Robert. *The Wild Places*. New York: Penguin Books, 2007.

———. *The Old Ways: A Journey on Foot*. New York: Penguin Books, 2012.

Machlis, Gary E., and Jonathan B. Jarvis. *The Future of Conservation in America: A Chart for Rough Water*. Chicago: University of Chicago Press, 2018.

Madley, Benjamin. *An American Genocide: The United States and the California Indian Catastrophe, 1846–1873*. New Haven: Yale University Press, 2017.

McPhee, John. *Encounters with the Archdruid*. New York: Farrar, Straus and Giroux, 1971.

———. *Basin and Range*. New York: Farrar, Straus and Giroux, 1981.

———. *Assembling California*. New York: Farrar, Straus and Giroux, 1993.

Moores, Eldridge M., Judith E. Moores, Marc C. Hoshovsky, Peter Schiffman, and Bob Schneider. *Exploring the Berryessa Region: A Geology, Nature, and History Tour*. Kneeland, CA: Backcountry Press, 2020.

Muir, John. *My First Summer in the Sierra*. Boston: Houghton Mifflin, 1916.

Olson, Sigurd F. *Lonely Land*. New York: Knopf, 2012.

Raphael, Ray, and Freeman House. *Two Peoples, One Place*. Eureka, CA: Humboldt County Historical Society, 2011.

Reiger, John F. *American Sportsmen and the Origins of Conservation.* New York: Winchester Press, 1975.

Robinson, Kim Stanley. *The High Sierra: A Love Story*. New York: Hachette Book Group, 2022.

Sayre, Nathan Freeman. *Working Wilderness: The Malpai Borderlands Group and the Future of the Western Range*. Tucson: Rio Nuevo Publishers, 2005.

Schweber, Nate. *This America of Ours: Bernard and Avis Devoto and the Forgotten Fight to Save the Wild*. New York: Mariner Books, 2022.

Snyder, Gary. *The Practice of the Wild*. San Francisco: North Point Press, 1990.

Stegner, Wallace. *Beyond the Hundredth Meridian: John Wesley Powell and the Second Opening of the West*. Lincoln: University of Nebraska Press, 1982.

———. *The Sound of Mountain Water: The Changing American West*. New York: Vintage Books, 2017.

Wheat, Frank. *California Desert Miracle: The Fight for Desert Parks and Wilderness*. San Diego: Sunbelt Publications, 1999.

Williams, Terry Tempest. *Refuge: An Unnatural History of Family and Place*. New York: Pantheon Books, 1991.

———. *The Hour of Land: A Personal Topography of America's National Parks.* New York: Farrar, Straus and Giroux, 2016.

Wilson, Randall K. *America's Public Lands: From Yellowstone to Smokey Bear and Beyond*. Lanham, MD: Rowman and Littlefield, 2014.

Zahniser, Howard. *The Wilderness Writings of Howard Zahniser*. Edited by Mark W. T. Harvey. Seattle: University of Washington Press, 2014.

Zaslowsky, Dyan, and T. H. Watkins. *These American Lands: Parks, Wilderness, and the Public Lands.* Washington, DC: Island Press, 1994.

# ABOUT THE AUTHOR

*Photo by Asher Moss*

*Josh Jackson* is a writer, photographer, and leading voice for public lands managed by the Bureau of Land Management (BLM). Through his evocative Forgotten Lands Project, Josh employs immersive storytelling and striking visual narratives to inspire appreciation and engagement with our least understood, least protected, and largely unknown landscapes. His advocacy work has been featured by the *Los Angeles Times*, *SFGate*, and the *Nature's Archive* podcast. He lives in Los Angeles with his wife and three children. *The Enduring Wild: A Journey into California's Public Lands* is his first book. Explore more of his work at forgottenlandsproject.com.

# A NOTE ON TYPE

This book is set in Joly Text, designed by Léon Hughes and released by Blaze Type in 2021. Joly draws on Dutch type styles from the eighteenth century. The headers are set in Adobe Garamond, a digital interpretation of one of the most widely loved typeface styles in typographic history.